Space

A TEMPLAR BOOK
First published in the UK in 2007 by Templar Publishing,
An imprint of The Templar Company plc,
The Granary, North Street,
Dorking,
Surrey,
RH4 1DN
www.templarco.co.uk

Conceived and produced by Weldon Owen Pty Ltd
61 Victoria Street, McMahons Point
Sydney, NSW 2060, Australia

Copyright © 2007 Weldon Owen Inc.
First printed 2008

Group Chief Executive Officer John Owen
President and Chief Executive Officer Terry Newell
Publisher Sheena Coupe
Creative Director Sue Burk
Concept Development John Bull, The Book Design Company
Editorial Coordinator Mike Crowton
Vice President, International Sales Stuart Laurence
Vice President, Sales and New Business Development Amy Kaneko
Vice President Sales, Asia and Latin America Dawn Low
Administrator, International Sales Kristine Ravn

Project Editor Jessica Cox
Designer Kathryn Morgan

ISBN: 978-1-84011-712-7

Colour reproduction by Chroma Graphics (Overseas) Pte Ltd
Printed by SNP Leefung Printers Ltd
Manufactured in China

10 9 8 7 6 5 4 3 2 1

A WELDON OWEN PRODUCTION

▶in siders

Space

Alan Dyer

templar publishing

introducing

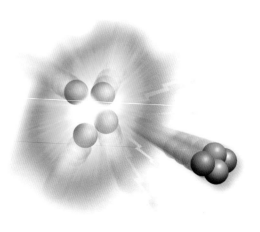

in *focus*

The Solar System

Stars and Galaxies

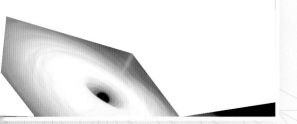

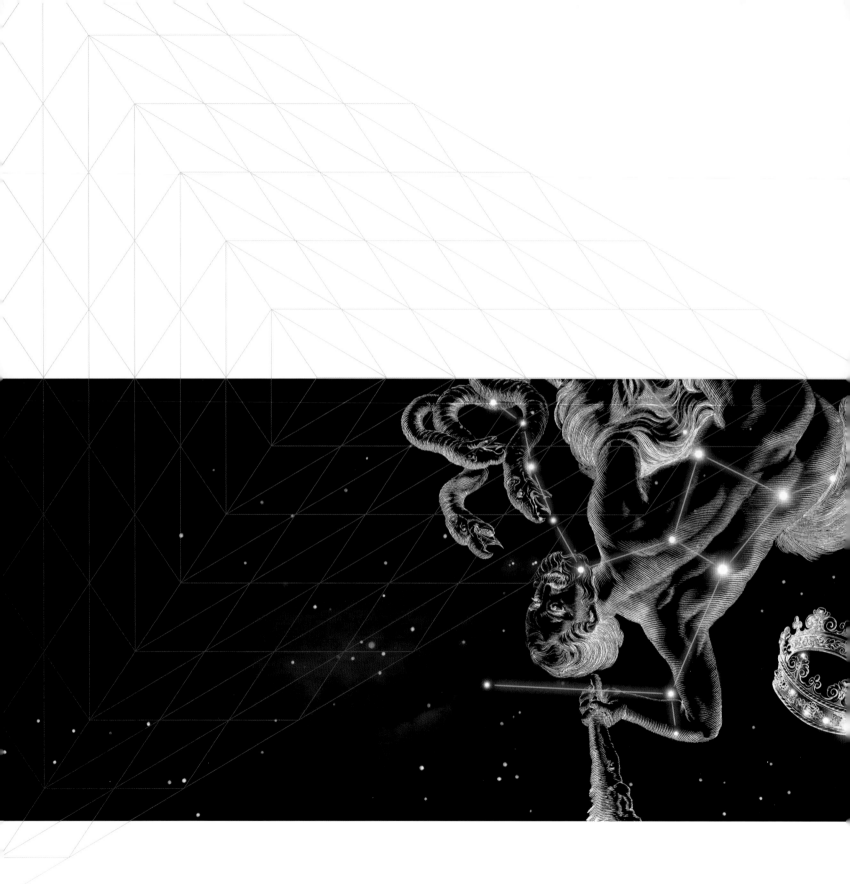

in*troducing*

Our Place in
Space

Space is really big! The planet we live on, Earth, is just one in a family of eight planets in our Solar System. Earth orbits a yellow dwarf star—the Sun. Though important to us, our Sun is just one of billions of other suns that make up a spiral-shaped galaxy called the Milky Way. We see our Sun in the day; our galaxy's other suns appear as the stars of the night sky. As big as our galaxy is, it is just one member of a local group of several dozen galaxies that, in turn, are just a small part of a vast network of billions of other galaxies. All these planets, stars and galaxies make up what we call the universe.

The cosmic neighbourhood

Our nearest neighbours in space are the worlds of our Solar System. They are close enough for people and robot probes to visit them. But beyond our neighbourhood, the vast distances of space make it impossible for us to travel to other stars and galaxies.

The Solar System *Zoom out, and Earth is just one planet in a system of worlds, which includes seven other major planets and several dwarf planets. It takes about six hours for light to travel from the Sun to the dwarf planet Pluto, on the outer edges of the Solar System.*

The blue marble
Astronauts who visited the Moon saw the distant Earth as a blue marble floating in the blackness of space, above the grey, lifeless landscape of the Moon.

Earth *Our home is a small, watery, blue planet that orbits about eight light-minutes from the warmth and light of an ordinary yellow star, the Sun. So far, Earth is the only place in the universe where we know life exists.*

The Local Group *A beam of light would take 2.5 million years to reach the closest large galaxy to ours, the Andromeda galaxy. This is one of our neighbours in a small family of galaxies called the Local Group.*

The Milky Way galaxy *Even zipping at the speed of light (300,000 kilometres per second/ 186,000 mi/s)—the fastest that anything can travel—a light beam would still take 100,000 years to travel across our galaxy.*

The known universe
To reach the most distant objects in the known universe, even travelling at the speed of light, it would take nearly 14 billion years. Along the way we would pass chains and clusters of billions of other galaxies.

LIGHT-YEARS

Light takes eight minutes to travel from the Sun to Earth. In one year light can travel almost 10 trillion kilometres (6 trillion mi), a distance called one light-year. So "light-year" measures distance, not time.

Light: 8 minutes

Sun

Earth

Car: 180 years (travelling at 60 kilometres per hour/40 mph).

Light-years from Earth to . . .

The Moon	1 light-second
The Sun	8 light-minutes
Pluto	6 light-hours
Proxima Centauri (nearest star)	4.2 light-years (ly)
Orion Arm of Milky Way	5,000 ly
Andromeda (nearest big galaxy)	2.5 million ly
Edge of visible universe	13.7 billion ly

It Started with a
Big Bang

The biggest question we can ask is—"How did it all begin?" How did the universe become the huge volume of space we see today, filled with billions of galaxies? Giant telescopes provide the clues—they show us that all the galaxies appear to be rushing away from one another. The universe is getting bigger and bigger, expanding like an inflating cosmic balloon. But if space is still getting bigger now, the universe must have been much smaller in the past. About 13.7 billion years ago the universe was packed into a tiny point smaller than an atom. This exploded in what astronomers call the Big Bang.

Bang! *A titanic explosion formed all space, energy and matter.*

First three minutes *The young universe was just atomic particles (such as electrons and protons), zipping about too fast to stick together to make atoms.*

A brief history of time

Space and time began in a superhot, superdense flash of energy. In a brief instant, the universe blew up in size. The force of that explosion still continues, expanding the universe we see around us.

Gas clouds form *As the universe expanded, it cooled. Like raindrops condensing within a cloud, atoms of hydrogen and helium formed into clouds of gas.*

First stars form *Gravity pulled together lumps in the gas to create the first stars. These giants soon exploded, seeding the universe with the ingredients for life, such as oxygen and carbon.*

FUTURE OF THE UNIVERSE

What is the fate of the universe? A mysterious force called "dark energy" may be growing in strength, driving a new, faster expansion of space.

The Big Rip
Will dark energy cause the universe to expand so fast even atoms eventually tear themselves apart?

The Big Chill
Or will the universe expand more slowly, until all stars burn out, chilling space to a frozen emptiness?

The Big Crunch
One idea was that the universe might crunch back together into a tiny point. This is unlikely.

The Big Bang

Today

The Big Rip

The Big Chill

The Big Crunch

First galaxies form
Small gas clouds collided and melded together, building up galaxies of billions of stars. Our Milky Way formed at this time.

We are here!
Today, we have discovered that not only is the universe still expanding, it is actually picking up speed.

Solar System forms *Billions of years later, a small cloud in a spiral arm of our galaxy collapsed into a spinning disk of gas and dust that created our Sun and its family of planets.*

The scale of time

What if the entire history of the universe took just one year? All of human history would then occur in the last few minutes.

1 Big Bang
Midnight, January 1

2 Gas clouds form
12:10 am, January 1

3 First stars form
January 5

4 Milky Way takes shape
Mid-January

5 Solar System forms
September 1

6 First life on Earth
September 22

7 Mammals appear
December 25

8 Humans appear
11:47 pm, December 31

9 Today
Midnight, December 31

January February March April May June July August September October November December

How the Solar System formed

The Solar System was born five billion years ago from a nebula—a cloud of dust and gas. The hot centre of this nebula, shrinking and spinning, formed the Sun. Leftover dust and gas formed planets around the Sun—the warm, rocky planets close in, the gas and ice planets farther out.

Nebula collapse A slowly turning nebula starts to shrink, heating up and flattening into a disk.

Proto-Sun The proto-Sun glows at the centre of the spinning disk.

Planets form Dust and gas clump together in the disk. These clumps collide to make larger planets.

The Sun's collapse The Sun will eventually expand and blast off a nebula. This will vaporise the inner planets.

Solar System now The Sun shines as a normal star. Its radiation blows away leftover dust and gas.

Held by the Sun

The Sun's gravity holds the objects of the Solar System in place. Their speed around the Sun stops them from falling into it, while the pull of the Sun's gravity stops them from flying off into space. The planets have been circling the Sun in exactly the same orbits for billions of years.

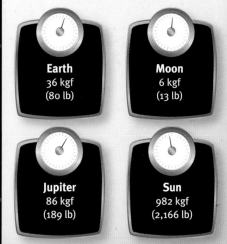
1. **Sun** Huge ball of hot gas at the centre of the Solar System

2. **Mercury** Airless, cratered, rocky planet; orbit: 88 Earth days

3. **Venus** Hot, rocky planet with thick atmosphere; orbit: 225 Earth days

4. **Earth** Rocky planet mostly covered by water; orbit: 365 days (1 year)

5. **Mars** Planet of red rock with a thin atmosphere; orbit: 687 Earth days

6. **Asteroids** Rocky objects between Mars and Jupiter, too small to be planets

7. **Ceres** Largest asteroid, smallest dwarf planet; orbit: 4.6 Earth years

8. **Jupiter** Stormy, spinning giant, a huge ball of gas; orbit: 12 Earth years

9. **Saturn** Brilliantly ringed planet, mostly gas; orbit: 29.5 Earth years

10. **Uranus** Ringed, icy planet tilted on its side; orbit: 84 Earth years

11. **Neptune** Windy, deep blue planet, made of ice; orbit: 165 Earth years

12. **Kuiper Belt objects** Tiny, icy worlds at the edge of the Solar System

13. **Pluto** Icy dwarf planet, second farthest from the Sun; orbit: 248 Earth years

14. **Eris** Dwarf planet bigger than Pluto, found in 2005; orbit: 557 Earth years

15. **Comets** Ice chunks that spin close to the Sun, warm up, and give off a tail

Comparing the
Planets

Planets are worlds too small and cold to "shine" like stars. Instead, they orbit around a star. The Solar System has eight major planets orbiting the Sun. Mercury, Venus, Earth and Mars orbit close in. When they formed, the Sun's intense heat kept them too warm to hold much ice or gas. So these "terrestrial planets" are made mostly of rock. The giants Jupiter, Saturn, Uranus and Neptune orbit much farther from the Sun. It is so cold where they formed, they held onto lots of ice and gas. Small dwarf planets are made of leftover rocky and icy debris.

DWARF PLANETS

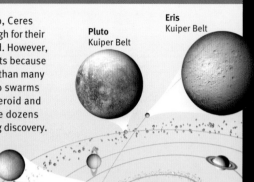

D warf planets like Pluto, Ceres and Eris are large enough for their gravity to make them round. However, they cannot be major planets because they are too small (smaller than many moons) and they belong to swarms of similar worlds—the asteroid and Kuiper belts. There may be dozens more dwarf planets awaiting discovery.

Pluto
Kuiper Belt

Eris
Kuiper Belt

Ceres
Asteroid belt

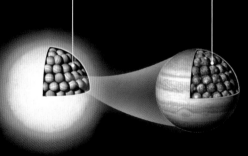

Weighing up the planets
Jupiter is the heavyweight of the Solar System. It weighs more than twice as much as all the other planets together.

Distance from the Sun
The small inner planets huddle close to the Sun's warmth. The cold outer planets spread far apart, with huge gaps between their orbits.

How big is the Sun?
As massive as Jupiter is, it is puny compared to the Sun's huge diameter of 1,392,530 kilometres (865,278 miles). More than 900 Jupiters could squeeze inside the Sun.

900 Jupiters
(or 1.3 million Earths)
could fit inside the Sun.

1,400 Earths
could fit inside
Jupiter.

Sun

Mercury 58 million kilometres
(36 million mi) from the Sun

Venus 108 million kilometres
(67 million mi) from the Sun

Earth 149 million kilometres
(93 million mi) from the Sun

Mars 228 million kilometres
(142 million mi) from
the Sun

Jupiter 778 million kilometres
(484 million mi) from the Sun

Saturn 1,427 million kilometres
(887 million mi) from the Sun

Saturn 120,540 kilometres
(74,900 mi) in diameter;
second biggest planet

Uranus 51,120 kilometres
(31,764 mi) in diameter;
third biggest planet

How many moons?
Generally, the bigger the planet, the more moons it has. Jupiter, the biggest planet, with the strongest gravity, has the most. Sometimes, astronomers discover new moons.

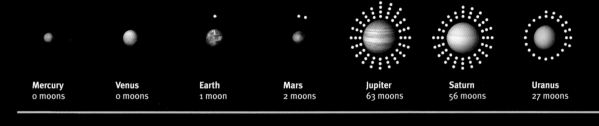

Mercury	Venus	Earth	Mars	Jupiter	Saturn	Uranus	Neptune
0 moons	0 moons	1 moon	2 moons	63 moons	56 moons	27 moons	13 moons

Jupiter
142,980 kilometres (88,844 mi) in diameter; biggest planet

How big are the planets?

The eight major planets come in two main sizes: gas giants and small, rocky planets. Jupiter is the biggest gas giant, while Earth is the biggest rocky planet. The dwarf planets, including Pluto, are so small that in this illustration they would not be much bigger than the dots on the end of each pointer.

HOT AND COLD

Venus 464°C (872°F)
Mercury 452°C (845°F) (day side)

Earth 15°C (59°F) (average)

Mars −63°C (−81°F) (typical day)
Jupiter −108°C (−162°F)
Saturn −139°C (−218°F)

Neptune −197°C (−323°F)
Uranus −201°C (−330°F)

Planets closer to the Sun, such as Venus and Mercury, are hotter than those farther out, like frigid Neptune. Gas giants have no solid surface, so astronomers measure the temperature of their cloudtops.

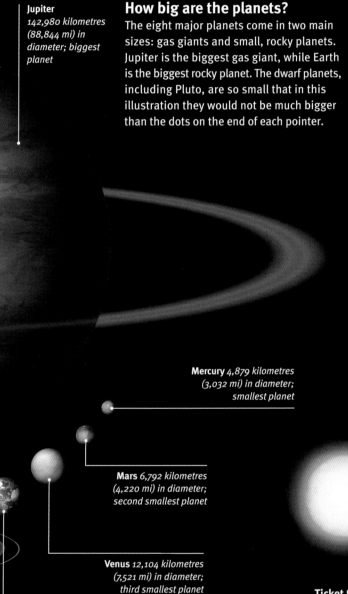

Mercury *4,879 kilometres (3,032 mi) in diameter; smallest planet*

Mars *6,792 kilometres (4,220 mi) in diameter; second smallest planet*

Venus *12,104 kilometres (7,521 mi) in diameter; third smallest planet*

Sun to Earth 20 years

Sun to Pluto 700 years

Ticket to Pluto

Distances in the Solar System are vast. It takes about five hours to fly from Los Angeles to New York. A fast passenger jet would take 20 years to travel from the Sun to Earth, and 700 years to get to Pluto!

Neptune *49,530 km (30,776 mi) in diameter; fourth biggest planet*

Earth *12,756 km (7,926 mi) in diameter; fifth biggest planet*

Uranus 2,871 million kilometres (1,784 million mi) from the Sun

Neptune 4,497 million kilometres (2,794 million mi) from the Sun

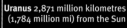

Snowballs in Space

Along with the planets, millions of smaller objects orbit the Sun. Asteroids are mostly rock and metal. Hundreds of thousands of them orbit between Mars and Jupiter in the asteroid belt. Meteoroids are even smaller bits of rock debris floating through space. There are other small objects, made mostly of ice, that usually lie far beyond Pluto. Some of these icy lumps can be forced close to the Sun, where the heat turns them into comets. Astronomers are also discovering that some asteroid belt objects are more ice than rock, which means comets and asteroids may be more alike than we once thought.

Blazing a trail

A comet is an icy object that gravity or a collision pushes close to the Sun. At its heart is a chunk of ice the size of a mountain. This icy mountain erupts with activity, giving off a cloud of gas and dust. Sunlight and solar particles blow this away in flowing tails that always point away from the Sun.

Dust tail *Solar radiation forces dust particles from the nucleus in a curving dust tail. This shines yellow from reflected sunlight.*

Deep Impact

In July 2005, the spacecraft Deep Impact reached a comet, Tempel 1. It shot a probe the size of a washing machine, called an "impactor", at the nucleus. The impact blew out a crater and a plume of water and dust molecules that had been deep inside the comet for billions of years. Deep Impact then studied the matter thrown out by the blast

Coma *Streaming gas and dust shoot from the coma, often millions of kilometres across.*

Deep Impact releases the impactor at the comet's nucleus.

Gas tail *Gas blown away by the solar wind forms a straight, bluish gas tail—the ion tail. Its colour comes from the glow of carbon monoxide gas.*

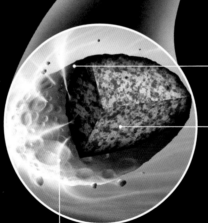

Surface *This is covered with a dark, dusty crust.*

Inside *The ice is mixed with rock and dust, like a huge, loosely packed, dirty snowball.*

Jets *Plumes of water vapour and dust erupt from surface cracks.*

A sign in the sky

In times past, people feared comets as omens of disaster. When Halley's Comet appeared in 1066, people saw it as a sign of King Harold of England's defeat by William the Conqueror.

The impactor hits at 37,000 kilometres per hour (23,000 mph).

Rocks in the Solar System

Most asteroids orbit in the asteroid belt between Mars and Jupiter. The Kuiper Belt, beyond Pluto, holds thousands of icy objects. Thousands more icy bits orbit even farther out, in the vast, round Oort Cloud. Some of these turn into comets when forced close to the Sun.

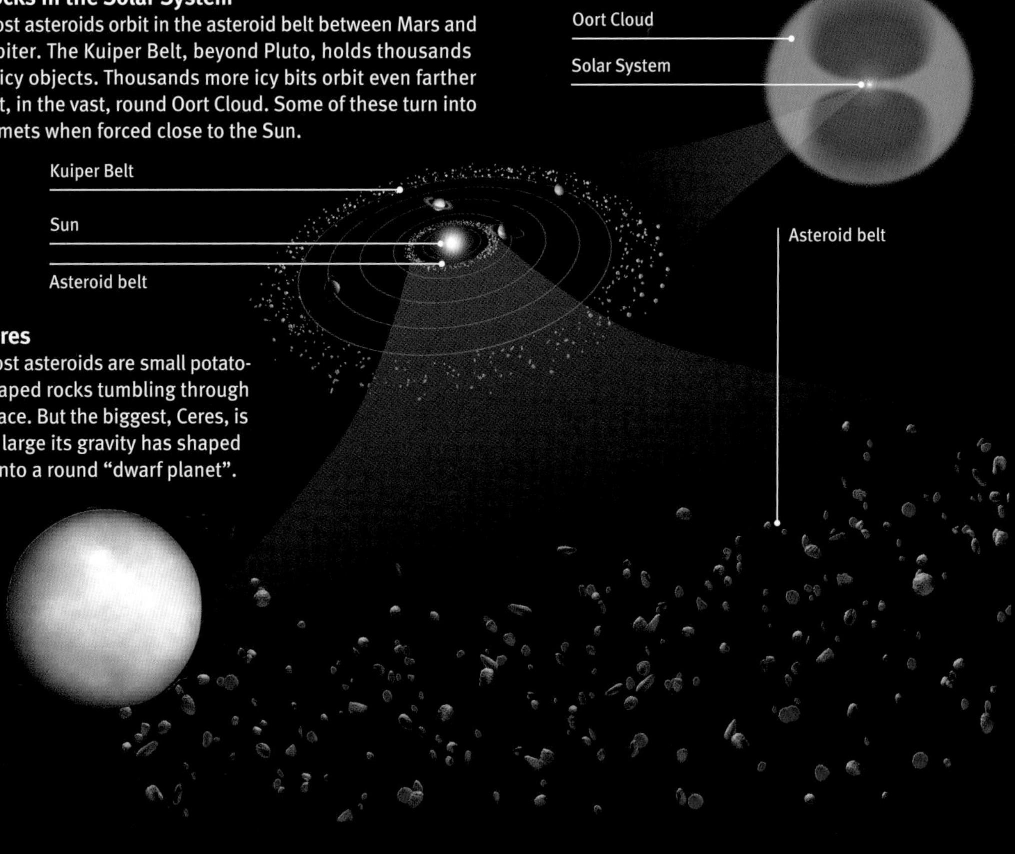

Oort Cloud

Solar System

Kuiper Belt

Sun

Asteroid belt

Ceres

Most asteroids are small potato-shaped rocks tumbling through space. But the biggest, Ceres, is so large its gravity has shaped it into a round "dwarf planet".

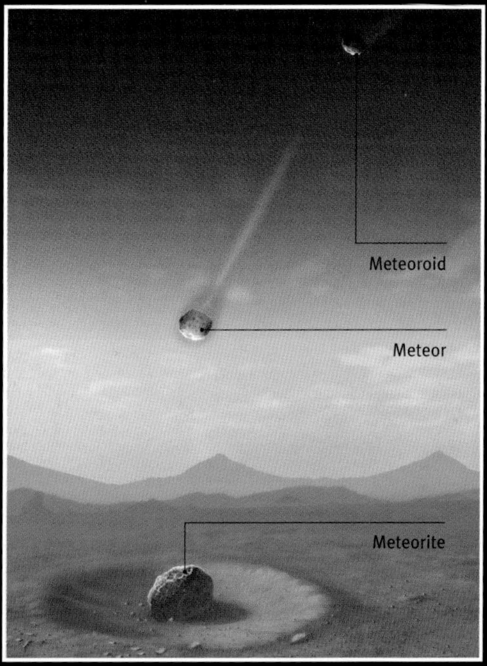

Asteroid belt

Meteoroid

Meteor

Meteorite

Streaking to Earth

Earth is constantly being hit by space debris. Comet dust burning up in our atmosphere creates meteors, or "shooting stars"—streaks of light in the night sky. Larger pieces of rock from the asteroid belt drift through space as meteoroids. If they survive the fall to Earth, they hit as meteorites.

Stars and Nebulas

...re giant balls of gas that, unlike planets, emit heat
...t. Our average yellow Sun is a typical star—not
too small or large, or too hot or cool. Red stars are the
coolest, while blue-white stars burn intensely hot. Stars
...om rotating clouds of cold gas that shrink in
...der the pull of gravity to create hot, shining
...which burn hydrogen gas. Some stars form
...e, single stars, though many form as pairs
...plets of stars, revolving around each other.
...n stars begin to run out of hydrogen fuel,
th... balloon in size. Some explode to form
an expanding cloud of gas, which may be
recycled into a new generation of stars.

Life cycle of stars

Stars have lives—they are born from
nebulas, then live shining as normal,
well-behaved stars, before inflating in
old age to become bloated red giants.
...fter that, stars smaller than our Sun
...de away, but stars much larger
...ie with a supernova bang.

1   **Massive giant star** *Betelguese,*
    in the constellation Orion

2   **Average-size star** *The Sun*

3   **Small red dwarf** *Proxima*
    Centauri, the closest star to us

Collapsing nebula
The pull of gravity causes
swirling clouds of gas to
shrink. As they get smaller,
the clouds heat up.

Young blue giant If a star
forms with lots of gas, it
becomes a blue giant.
This burns many times
hotter than the Sun.

Supernova When the
supergiant suddenly runs out
of fuel in its hot core, the core
collapses. A powerful shock
wave rushes outwards, blowing
off the star's outer layers.

Supergiant After just
a few million years,
the blue giant
balloons to become
an even larger red
supergiant.

Stars come in all sizes. The biggest and brightest,
called supergiants, could swallow all the planets
out to Mars. The smallest, white dwarfs, are no
bigger than Earth. Red dwarfs, though small and
dim, are the most common—there are hundreds
of red dwarfs for every red giant.

Yellow dwarfs
Stars like our Sun

Red dwarfs
Smaller stars
than our Sun

Earth
Same size as
white dwarfs

Red giants
Ageing, bloated stars

Blue giants
Stars containing
a lot of gas

Yellow dwarf
At same scale
as giant stars

Supergiants Old stars that might explode

Young yellow star After it forms, a star the size of the Sun shines as a stable yellow dwarf. It dependably pours out heat and light for billions of years.

Middle-aged star A star that forms with less gas than the Sun becomes a small, cool red dwarf. This burns its fuel slowly, quietly shining for tens of billions of years.

Ageing red dwarf Eventually, even a red dwarf will bloat in size to become a modest giant. This process takes so long no red dwarf may have reached this stage.

Fading star The red dwarf will eventually fade in brightness and shrink in size. It becomes a dim white dwarf or perhaps a burnt-out black dwarf.

Red giant As it runs out of fuel, the Sun expands to a red giant, engulfing the inner planets and burning them to cinders.

Planetary nebula Five billion years from now, the Sun puffs off its outer layers, creating expanding shells of hot gas called a planetary nebula.

Stellar remnant The remaining core shrinks under gravity, forming a dense neutron star or black hole, an object too dense for even light to escape.

White dwarf The exposed core contracts to become a superhot white dwarf star. This can continue to shine for billions more years.

Kinds of nebulas

Dark nebula Some nebulas are so dust-filled they appear like dark clouds hiding what lies within or behind them.

Emission nebula Hot, newborn stars can heat up the nebulas surrounding them, causing the gas to glow in beautiful colours.

Reflection nebula Some cold, dusty nebulas don't shine with their own light, but instead reflect light from nearby stars.

Planetary nebula Ageing, Sun-like stars blow off shells of gas that can form complex spirals, rings and disks.

Supernova nebula Massive stars blow themselves apart, leaving behind a glowing cloud of debris around them.

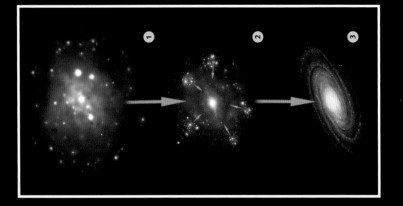

A Universe of
Galaxies

Most stars in the universe belong to collections of billions of stars called galaxies. Our Sun belongs to the Milky Way galaxy. Through telescopes, countless other galaxies can be seen, some like the Milky Way, but others very different in shape and size. The entire visible universe contains at least 100 billion galaxies, each with billions of stars and nebulas. Though new stars are always forming within galaxies, the galaxies themselves are ancient. Most formed not long after the origin of the universe in the Big Bang. Galaxies group together into clusters, some with hundreds of members.

How galaxies are born

Galaxies are giant structures, probably built up from small, irregular "proto-galaxies" that merged to create larger ones. It is still a mystery exactly how galaxies formed because they did so billions of years ago.

1 Proto-galaxies form
Soon after the Big Bang, clumps of gas began to shrink to form proto-galaxies made of hot, massive stars, the first to form in the universe.

2 Proto-galaxies merge
These ragged young galaxies, made up of stars and gas clouds, came together under gravity and merged to form larger, well-formed galaxies.

3 Modern-day galaxy
Some of these early galaxies spun and flattened to form spiral galaxies. Some spirals collided and merged again to form giant ellipticals.

Kinds of galaxies

Like stars, galaxies are grouped by size. Dwarf galaxies can be 100 times smaller than the Milky Way, with only a million stars. Giants ten times the size of our galaxy can contain a trillion stars. Unlike stars, galaxies come in many shapes.

Looking back in time

The Hubble Space Telescope stared at a tiny region of sky for 11 days, revealing young galaxies near the edge of the visible universe. These were born shortly after the Big Bang.

Elliptical galaxy Some of the largest galaxies are featureless round globes or football-shaped ellipses. Stars orbit in all directions, like bees buzzing around a hive.

Lenticular galaxy *These lens-shaped galaxies look much like elliptical galaxies, but are surrounded by a flat disk of stars that resembles a spiral galaxy, without the prominent arms.*

Barred spiral galaxy *These galaxies have two large arms twisting out from the ends of a straight bar of stars, which is attached to a small bright core. The Milky Way is a barred spiral.*

Galaxy cluster These can be small, like our Milky Way's Local Group. Some vast clusters contain thousands of galaxies.

Globular star cluster Globes of hundreds of thousands of stars, usually satellites of galaxies; often made of old, unchanging stars.

Open star cluster Loose collections of stars, found in the arms of spiral galaxies; often contain young stars.

STAR AND GALAXY CLUSTERS

S tars often form in clusters of dozens to thousands of stars. These are found in and around galaxies, grand collections of billions of stars. Galaxies also come in clusters; most galaxies belong to a cluster.

Irregular galaxy *Some galaxies have no distinct shapes. They are ragged collections of millions of stars, mixed with large nebulas where new stars are forming.*

Spiral galaxy *In these galaxies, several arms spiral around a dense core. The core contains older yellow stars, while the arms are filled with young, hot blue stars.*

Constellations

Since ancient times, people have looked to the sky and imagined the stars joined together to make patterns, called constellations. By learning to recognise these patterns, people could find their way at night. Stories about mythical people and animals were retold through the ages—the sky served as a great teacher. Many of the constellations we use today were invented 4,000 years ago in the Middle East, handed down to us from the Greek and Roman civilisations 2,000 years ago.

Actual positions of stars

Pattern we see in our sky

Our view of the sky

Earth

How far are the stars?

It looks as if the stars in a constellation like Orion are all the same distance from us, but in reality some stars are close while others are very far away. The pattern they make in our sky is just an illusion. Elsewhere in our galaxy, aliens would see the same stars forming a very different pattern in their sky.

Hercules *A son of Zeus, this legendary strongman of Greek mythology was ordered to perform 12 impossible tasks. Among them was battling monsters such as the many-headed Hydra, whose central head was immortal.*

Corona Borealis *This circlet of stars is the jewelled crown worn by the god Dionysus. He tossed his crown into the sky as a sign of his love for the mortal Ariadne, daughter of Minos, king of Crete.*

Same stars, different forms

People around the world all see the same stars, but imagine them differently in a way that is meaningful to their culture. The stars provide a join-the-dots game that lets us draw what we like, then tell wonderful stories about what we imagine in the sky.

Orion To the Greeks, these stars were Orion, a great hunter who boasted he could kill all the animals on Earth. As punishment, the Earth goddess had him killed by a scorpion sting from Scorpius.

Ancient figures

It takes a lot of imagination to see bears and heroes in the sky. In an old star atlas published in 1690, Johan Hevelius drew these fanciful depictions of Ursa Major—the Great Bear—and other constellations of the northern sky. Today, we often join the dots to make simpler patterns, like the Big Dipper.

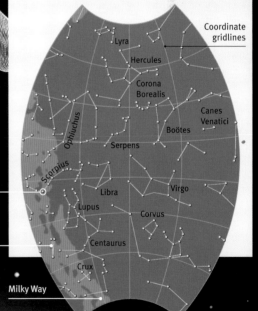

The Big Dipper A wagon or plough in Europe, in North America people see these seven stars as a pot or drinking ladle.

Ursa Major *Many northern cultures saw these stars as a bear. To the Greeks, these stars were the beautiful Callisto, who was turned into a bear by the jealous goddess Hera. Her son, Arcas, is the smaller bear, Ursa Minor.*

Boötes *These stars represent the bear driver, Boötes. With his two hunting dogs, Canes Venatici, Boötes herds the two bears around the northern sky. In Greek legend, Boötes is also credited with inventing the plough.*

MAPPING THE SKY

Like maps of the world, sky maps chart the positions of stars and constellations against a grid of coordinate lines, similar to latitude and longitude. This map shows stars visible April to July, but different stars appear throughout the year as Earth orbits the Sun.

Coordinate gridlines

Lyra

Hercules

Corona Borealis

Canes Venatici

Ophiuchus

Boötes

Serpens

Scorpius

Stars

Libra

Virgo

Lupus

Corvus

Lines defining constellations

Centaurus

Crux

Milky Way

Kimono sleeve In Japan, they are sometimes called *Sode Boshi*, "kimono sleeve stars". They look like the long sleeves of a woman's kimono dress, flowing gracefully down to the southern sky.

Osiris In ancient Egypt, the stars were Osiris, god of light. His evil brother, Set, god of darkness, murdered Osiris. But the goddess Isis immortalised Osiris as a celestial mummy in the sky.

Gazing into Space

People have looked up at the stars for thousands of years. Ancient civilisations used the sky as their calendar and clock, tracking the motions of the Sun, Moon and stars. Because the sky appeared to be turning around us, people believed Earth was the centre of the universe. The history of astronomy has been a series of discoveries proving just how wrong this picture was. Since Galileo first used a telescope 400 years ago, bigger and better telescopes have peered farther into space, revealing a universe much larger than we ever imagined. What will we discover in the future? We can say only that it will be amazing.

Antu (the Sun) *The four main telescopes have names from the local Mapuche language. Antu, the first VLT completed, is named for the Sun.*

Dome shutters These slide open at night to 10 metres (33 ft), so the telescope can see out.

Yepun (Venus) *By combining the light from all four identical siblings, the VLT works like one giant telescope.*

Other solar systems

Our Solar System is not alone in the universe. Using telescopes like the VLT, astronomers have detected other suns with orbiting planets. The telltale sign: the light from these suns shifts colour. The unknown planet's gravity tugs its sun (the star) towards us, so it looks bluer. Then it tugs the sun away from us, making it seem redder.

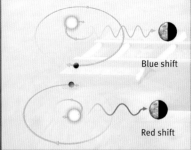

Blue shift

Red shift

Large main mirror Each telescope collects light using a dish-shaped mirror 8.2 metres (27 ft) across. This reflects light up to the top, secondary mirror.

Secondary mirror The small mirror at the telescope's top shoots the light back down to instruments and cameras at the telescope's bottom.

A timeline of astronomy

On many dates in our history, new ideas, inventions and discoveries changed our ideas about the universe.

2000 BC
Stonehenge, possibly an ancient observatory, is erected in England.

150 BC
Hipparchus of Nicea measures length of the year, distances to the Moon and Sun, and compiles first star catalogue.

1609
Galileo Galilei is the first person to use a telescope for astronomy. He discovers craters on the Moon and the large moons of Jupiter.

32,000 BC
Stone-age people mark bones with phases of the Moon.

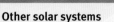

350 BC
Aristotle claims that Earth is centre of the universe.

1543
Nicolaus Copernicus publishes his theory of a Sun-centred universe on his deathbed.

1619
Johannes Kepler completes his laws of planetary motion.

1687
Isaac Newton publishes his law of gravity.

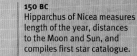

Kueyen (the Moon) *Each VLT has cameras and instruments for special observations. Kueyen has an instrument that can see ultraviolet light.*

Melipal (the Southern Cross) *This has an instrument to see infrared light. Another can analyse light from dozens of galaxies at once.*

Telescope mount The mirrors are held in a motorised mount that can swing anywhere in the sky, then track objects across the sky all night.

Altitude axis Each mount can rotate up and down to aim the telescope anywhere from the horizon to straight up.

Azimuth axis Each mount can spin in a circle to aim the telescope at any compass point in the sky.

Fixed platform Each telescope and rotating enclosure is mounted on a fixed concrete foundation, solidly dug into the ground to prevent any vibrations.

Airflow louvres Air flows through vents to keep the temperature inside the same as outside. This prevents blurring heat waves.

DISCOVERING THE SKIES

Ideas about the universe are always changing. When we discover more distant planets, or surprising objects like white dwarfs and quasars, we have to revise our theories of how the universe works.

Early observations
On 4 July 1054, Chinese skywatchers saw a brilliant "new star" appear. Their records provide us with valuable data about this exploding star, a supernova. The Crab Nebula has resulted from this.

Firsts in the heavens

165 BC	Chinese first observe sunspots
1054	Chinese witness supernova
1066	Halley's Comet is recorded in Bayeux Tapestry
1781	Sir William Herschel discovers Uranus
1915	Sirius B becomes the first known white dwarf
1930	Clyde Tombaugh discovers Pluto
1932	Cosmic radio waves detected
1963	First quasar found
1967	Pulsars discovered
1972	First black hole identified
1995	First extrasolar planet discovered
2006	Pluto and other KBOs reclassified

Dwarf planets These worlds (Eris, Pluto and Ceres) are all much smaller than Earth.

Earth

Eris

Pluto

Ceres

The Very Large Telescope

Simply called the Very Large Telescope (VLT), these four giants are amongst the world's biggest telescopes. While they were built by European astronomers, they are in northern Chile, South America, where the sky is clear and it rarely rains. Each telescope has instruments to take pictures or split the light into its colours.

1785
William Herschel publishes the first map of our galaxy.

1838
Friedrich Bessel first measures the distance to a star.

1905
Albert Einstein publishes *The Special Theory of Relativity*, followed 10 years later by *The General Theory of Relativity*.

1920s
Edwin Hubble discovers that the universe is expanding.

1998
The universe's expansion is found to be accelerating.

2003
Age of universe measured to be 13.7 billion years old.

1845
Lord Rosse discovers the first spiral galaxy, the Whirlpool galaxy.

1908
Henrietta Leavitt discovers method for measuring cosmic distances.

1918
Harlow Shapley discovers we are not at the centre of our galaxy.

1965
Discovery of cosmic microwave background provides evidence for the Big Bang theory.

Exploring Space

We have learnt almost everything we know about the planets with the help of robots. We call them space probes. Since the 1960s, they have roamed the Solar System as our robot eyes and ears, to explore where no one has ever gone. People have travelled only as far as the Moon, but robots have visited all the major planets, and one is on its way to the dwarf planet Pluto. Some probes fly past their targets, then continue out of the Solar System, never to return. Others orbit their destination planets or land to explore the surface. A few probes return to Earth, bringing samples back—perhaps dust from a comet, or one day rocks from the surface of Mars.

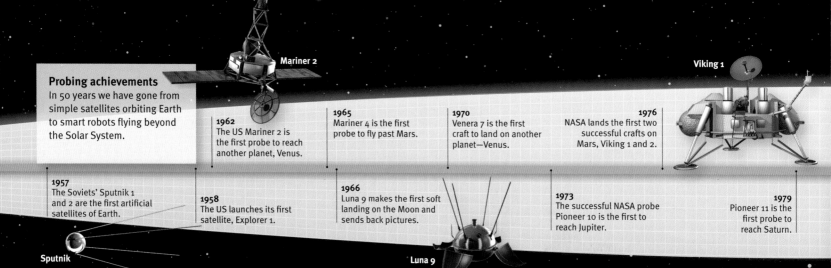

Launch *A powerful rocket propelled New Horizons away from Earth at 58,000 kilometres per hour (36,000 mph), faster than any other probe.*

Jupiter gravity assist *Only a year later, the little probe sped past Jupiter, getting a further boost in speed from the giant planet's gravity.*

Interplanetary cruise *For eight years it will sleep, signalling Earth just once a week, waking 50 days per year to make measurements.*

Perhaps Europan plant and animal life thrives in deep-sea vents, where heated water erupts from below.

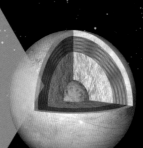

Possible life?
One day a probe may explore Europa, a moon of Jupiter that contains a global ocean under a frozen ice crust. Where there is liquid water, there may be life.

Travelling with New Horizons
Years in the planning, a probe was finally launched towards Pluto in January 2006. No probe has visited Pluto; fuzzy views from Earth show little on this tiny world. New Horizons will reveal details as small as football fields on Pluto and its moons.

Probing achievements
In 50 years we have gone from simple satellites orbiting Earth to smart robots flying beyond the Solar System.

Mariner 2

Viking 1

1962
The US Mariner 2 is the first probe to reach another planet, Venus.

1965
Mariner 4 is the first probe to fly past Mars.

1970
Venera 7 is the first craft to land on another planet—Venus.

1976
NASA lands the first two successful crafts on Mars, Viking 1 and 2.

1957
The Soviets' Sputnik 1 and 2 are the first artificial satellites of Earth.

1958
The US launches its first satellite, Explorer 1.

1966
Luna 9 makes the first soft landing on the Moon and sends back pictures.

1973
The successful NASA probe Pioneer 10 is the first to reach Jupiter.

1979
Pioneer 11 is the first probe to reach Saturn.

Sputnik

Luna 9

Telescope The probe will use this lens to take long-range and close-up images of Pluto and its moons.

Atmosphere detectors These will detect any gases and particles streaming away from Pluto's thin atmosphere.

Antennae New Horizons will never return. It transmits all its data back to Earth using these antennae.

M ars is the prime target for current and future missions. Probes are also exploring Mercury, Venus and Saturn. Scientists hope to launch probes to asteroids, comets and the moon Europa.

Venus Express
This European Space Agency (ESA) mission orbits Venus, charting its complex weather patterns, analysing its atmosphere and looking for evidence of active volcanoes.

Current missions

Cassini Saturn orbiter, launched 15 October 1997
Messenger Mercury mission launched 3 August 2004
Mars Reconnaissance orbiter launched August 2005; in orbit around Mars
Venus Express launched 9 November 2005; in orbit around Venus
New Horizons mission to Pluto and beyond; launched 19 January 2006

Piano-size probe
Though bigger than a 10-year-old (1.3 m/4.5 ft), New Horizons is small for a space probe.

Power source Travelling too far from the Sun for solar power, New Horizons gets electricity from nuclear-powered generators.

Camera This will allow the probe to capture visible and infrared maps of Pluto's and Charon's surfaces.

Steering rockets Side-mounted rockets turn and aim the piano-size probe.

Pluto–Charon encounter
The probe will finally reach its destination in July 2015, photographing Pluto and its moons.

Encounters in the Kuiper Belt *New Horizons will speed on, possibly to encounter one or more Pluto-like objects in the Kuiper Belt.*

Voyager 2

Giotto

1986
Voyager 2 reaches Uranus; the ESA's Giotto and the Soviets' Vega 1 and 2 are the first probes to intercept a comet—Halley's.

1989
Voyager 2 reaches Neptune.

1990
NASA's Magellan reaches Venus and radar-maps its entire surface.

1997
Pathfinder lands and deploys the first Martian rover, Sojourner.

2000
The probe NEAR–Shoemaker orbits and maps the asteroid Eros.

NEAR–Shoemaker

2004
Cassini–Huygens mission arrives at Saturn; Mars Rovers land on Mars.

2005
NASA's Deep Impact probe arrives at Comet Tempel 1; Huygens lands on Titan.

2006
New Horizons probe launched; NASA's Stardust returns comet dust samples.

Venturing into Space

Humans first ventured into space in the early 1960s, with short flights around Earth. From 1969 to 1972, the Apollo program landed people on the Moon for stays of only a few days. Since then, astronauts have stayed close to home, building Earth-orbiting space stations such as Skylab, Mir and the International Space Station (ISS). These projects have taught us how to live in space for many months, and how to build large structures in space. Now, plans are underway to send astronauts back to the Moon, this time to set up a permanent base, then use that as a departure point to go on to Mars.

Antennae *Dish antennae aimed at Earth would allow astronauts to send back experiment results to mission control and to talk to friends and family at home.*

Airlock node *Astronauts would put on their spacesuits here, then exit once air had been pumped out of the airlock, just as astronauts do now on the ISS.*

A home on the Moon
Probably the best place to set up a lunar base is near one of the Moon's poles, where sunlight for electricity is available most of the time. Water might also exist as buried ice that could be mined for astronauts to use.

Pressurised rover *Some rovers would be pressurised like a spaceship, allowing astronauts to live in them for many days or weeks without having to wear a spacesuit.*

Great milestones
The last 50 years of manned space exploration have been filled with historic milestones, both great achievements and tragic losses.

1961
Soviets fly the first man into space, Yuri Alekseyevich Gagarin.

1962
John Glenn is the first American to orbit Earth.

1963
Soviet cosmonaut Valentina Vladimirovna Tereshkova is the first woman in space.

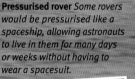

1965
Soviet cosmonaut Alexei Leonov performs the first "space walk".

1969
In July, Apollo 11 lands the first men on the Moon.

1971
Soviets launch the first successful space station, Salyut 1.

1972
Final Apollo mission to the lunar surface—no humans have been back since.

1981
Maiden flight of the first American Space Shuttle, Columbia.

Earth *From a base built at a lunar pole, Earth would appear just above the horizon. It would always remain in the same place in the sky, going through a monthly cycle of phases like the Moon does when seen from Earth.*

THE FUTURE OF SPACE

The US space agency, NASA, plans to return astronauts to the Moon by 2020. After establishing a lunar base, NASA may begin to send astronauts much farther into space, to land on Mars.

On to Mars
Because Mars is so far away, astronauts would need to stay for many months. To survive such a long stay, they would extract water and oxygen from the Martian soil.

Likely dates

2008 Unmanned Lunar Reconnaissance Orbiter launch
2010 Final flight of Space Shuttle; ISS complete
2014 First flight of crewed Orion spaceship to ISS
2020 First flight of Orion and astronauts to the Moon
2030 First astronauts to land on Mars

Orion Crew Vehicle
NASA is building Orion, a new spaceship to fly as many as six astronauts to the ISS and four to the Moon, for missions several months long.

Gangway *Returning from a moonwalk, astronauts could brush moondust off their suits and boots here, so they do not get grimy lunar dust on everything inside the base.*

Power storage *During times when the Sun is not visible, electricity could come from battery-like storage units charged up by solar panels.*

Astronaut at work *While some astronauts conduct scientific expeditions, others would work outside the base, fixing and maintaining life-support systems.*

Solar power units *Power would come from solar panels that convert sunlight into electricity. These slowly turn to follow the Sun around the sky during the four-week-long lunar "day".*

1986
Soviet Union launches its space station Mir; the Space Shuttle Challenger explodes shortly after lift-off, killing all seven astronauts on board.

1983
Sally Kristen Ride is the first American woman in space.

1988
Construction begins on the ISS.

2001 Aging and failing Mir space station is crashed into Earth.

2003
Americans lose another shuttle, Columbia, grounding the Space Shuttles for three years.

2004
Chinese launch first taikonaut (astronaut) into space.

The International
Space Station

Orbiting 390 kilometres (240 mi) above our heads is a giant laboratory where astronauts work in space, living for months without the pull of gravity. Still under construction, the International Space Station (ISS) is a project of five space agencies, representing 16 nations. When completed in 2010, the ISS will be as big as a football field, hosting a permanent crew of seven astronauts working in five laboratories. As the ISS orbits Earth, it teaches us how to work together to build structures in space, and about the effects of living in space so that astronauts can one day travel to the Moon and Mars.

Integrated Truss *A long girder anchored to the Destiny lab forms the backbone of the ISS. The huge solar panels attach to the truss.*

Past and present
Construction of the ISS began in 1998, and t will not be complete until at least 2010.

2006 Seven years later, the ISS had grown to be the largest structure built in space.

1999 When the station began, it consisted of just one Russian and one US module.

Station of many parts
The ISS is too big to launch in one piece. Instead, dozens of US Space Shuttle and Russian Proton rocket flights each take a section into orbit, for astronauts to assemble during spacewalks. This is how it looked mid-2006.

Solar arrays *These unfold like an accordion to become panels of solar cells that are as long as football fields. They supply electricity to the ISS.*

ROAD TO COMPLETION

When complete, the ISS will have more solar panels to power extra laboratories, room for seven astronauts and ports for docking European and Japanese cargo craft.

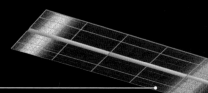

Completed
ISS

Zvedza Service Module *Added in 2000, this Russian module serves as the living quarters and the docking port for unmanned supply rockets.*

Nodes and nodules

2007	European Transfer Vehicle	2008	Fourth and final solar panel array
2007	US Node 2 Hub (utilities)	2008	Russian Multipurpose Lab
2007	European Columbus Lab	2009	Expanded crew quarters
2007	Canadian robot hand Dextre	2009	Japanese Transfer Vehicle
2007/08	Japanese Kibo Labs	2010	US Node 3 Hub and Cupola

Progress resupply vehicle *Every few months an unmanned Progress craft arrives at the ISS, bringing new supplies and taking rubbish away.*

Pirs Docking Compartmen *Russian cosmonauts doin spacewalks can exit and enter th ISS here. Soyuz craft bringing ne crew members also dock her*

Heat radiators *These panels allow extra heat generated by the ISS to escape into space, keeping the inside of the ISS a comfortable temperature.*

Canadarm *This Canadian-built robotic arm can move around the ISS to swing large components into position and to assist in repairs.*

Experiments Inside Destiny, astronauts discover how things, including plants and humans, behave in the zero-gravity environment of Earth's orbit.

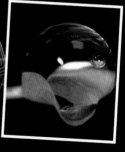

Quest Joint Airlock
This airlock allows astronauts to put on spacesuits and exit the ISS to work outside during spacewalks.

Soyuz capsule *Launched from Russia, Soyuz craft bring new crew members. The capsule is a lifeboat to allow astronauts to escape in an emergency.*

Space shield The exterior is made of bulletproof material to protect against meteorites.

Hanging about Astronauts float in and out of the lab in zero gravity.

Destiny lab

Launched in 2001, the US-built Destiny laboratory contains racks of equipment for zero gravity research, as well as systems for controlling the robot arm and a window for observing Earth.

Braving the Void
Spacesuit

When astronauts perform spacewalks, they need to wear their own "spaceship"—a flexible spacesuit that protects astronauts from the deadly vacuum of space. These suits provide oxygen to breathe and insulate astronauts from the searing heat (121°C/250°F) when they are in sunlight and brutal cold (-156°C/-250°F) in darkness. The thick layers of the suit also stop dangerous radiation and high-speed micro-meteorites. Without a working spacesuit, astronauts could not survive more than a few seconds in space. NASA spacesuits, called EMUs, have 11 layers of protective cloth and metal foil. On Earth, they weigh more than the astronaut. In orbit, they weigh nothing.

Communications *Astronauts wear a cap with earphones and a microphone for radio contact with other astronauts or mission control on Earth.*

Life-support backpack *The backpack supplies oxygen and electricity and removes poisonous carbon dioxide from the air inside the suit.*

Headlamp *Because they are in darkness on the night side of Earth for half of every 90-minute orbit, astronauts need helmet lights to keep working.*

Helmet and visor *Made of unbreakable plastic, the helmet has a visor coated with gold that blocks out dangerous ultraviolet light.*

Water bag *When astronauts get thirsty they suck on a straw beside their mouths for a drink of water from a plastic bag inside the suit.*

Hard upper torso *Suits are made in two halves that lock together with a metal collar. The upper half is made of hard fibreglass with flexible fabric arms.*

Display and control module *A front chest pack contains controls so astronauts can adjust the temperature, oxygen flow, lights and communication.*

Gloves *Thick gloves are custom made for each astronaut and include heaters to warm the palm and fingertips.*

Tools Vital tools are attached to the suit with tethers to keep them from floating off into space.

Lower torso Astronauts first put their legs into the lower half of the suit, then move up into the upper half. Other astronauts help connect the two halves together.

Absorption garment Astronauts cannot get out of the suit just to go to the bathroom. Instead they wear a space diaper—they just "go" in their suit!

Knee joint Flexible joints at the knees, ankles, waist, elbows and shoulders allow astronauts to bend in the otherwise stiff, pressurised suit.

Liquid cooling garment The underwear layer contains plastic tubes that circulate cooling water near the astronaut's skin.

Tethered to safety Astronauts can be tethered to their spacecraft by a cord. This is called an umbilical when it supplies astronauts with oxygen.

Space boots To be airtight, the boots are part of the legs. They have no treads—astronauts have no need for traction in zero gravity.

Keeping cool in space

Beneath the suit, astronauts wear a liquid cooling garment. The bottom layer has a network of tubes to circulate cool water to keep astronauts from overheating in the airtight suit.

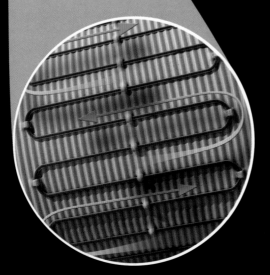

WORKING IN SPACE

Building the ISS, or fixing the Hubble Space Telescope, is hard work. Teams of two astronauts spend as many as seven hours outside at a time. Completing the job may take several spacewalks.

Making repairs To keep Hubble working, astronauts visit it every few years to repair faulty parts and install new, more sensitive cameras.

In training To simulate the zero gravity of space, astronauts practise their spacewalks on Earth by floating underwater in a giant swimming pool.

Out on a limb Astronauts often ride on the end of the robot arm, using it as a crane to lift them so they can work in hard-to-reach places.

Hubble Space Telescope

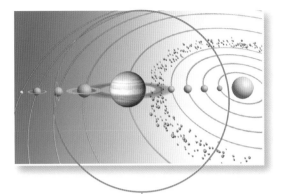

JUPITER: THE FACTS

ORIGIN OF NAME: King of Roman gods

DISCOVERED: Known since antiquity

DISTANCE FROM THE SUN: 779 million km (484 million mi)

VOLUME: 1,321 (Earth = 1)

GRAVITY AT CLOUDTOPS: 2.5 (Earth = 1)

DIAMETER: 142,980 km (88,844 miles)

CLOUDTOP TEMPERATURE: −110°C (−160°F)

NUMBER OF MOONS: 63

LENGTH OF DAY/YEAR: 9.9 hours/11.86 Earth years

Fast facts Fast facts at your fingertips give you essential information on each object.

Locator This diagram of the Solar System shows you exactly where the object is located. In the Stars and Galaxies section are photographs of each object.

Cross-section This cross-section shows inside each Solar System object. It stretches from the atmosphere to the core. This feature does not appear in the Stars and Galaxies section.

GAS

SURFACE

CONVECTION ZONE

RADIATION ZONE

CORE

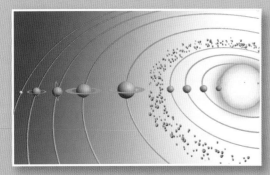

THE SUN: THE FACTS

ORIGIN OF NAME: From Old English word *sunne*

DISCOVERED: Known since antiquity

AGE OF THE SUN: 4.6 billion years

GRAVITY: 28 (Earth = 1)

VOLUME: 1,304,000 (Earth = 1)

DIAMETER: 1,392,530 km (865,278 mi)

CORE TEMPERATURE: 15 million°C (27 million°F)

NUMBER OF PLANETS/DWARF PLANETS: 8 planets/3 dwarfs

ROTATION PERIOD: 25.4 days (at equator)

Engine of the System

The Sun

The Sun is the centre of the Solar System. It is the energy source that lights and warms the planets. Without it, they would be dark worlds colder than Pluto, and life on Earth would be impossible. Like all stars, the Sun is a giant ball of superheated hydrogen gas, which is its fuel. The Sun has radiated enormous amounts of energy every second for almost 5 billion years, and has enough hydrogen to shine for 5 billion years more. Eventually, the Sun will become a red giant star, its size and heat burning the inner planets, including Earth, into cinders. It then will shrink into a feeble white star.

Powered by fusion

The Sun is not on fire, like wood burning. Instead, superhot hydrogen atoms smash together deep in the core to make heavier helium atoms and give off energy, in a process called nuclear fusion.

Hydrogen atoms are made of tiny single particles called protons.

Four protons collide to create one helium atom.

Helium emerges as two of the protons turn into neutrons.

Energy is given off as gamma rays.

Solar *towering ton as hot as te of degrees, either rise up from or rain down onto the Sun's surface.*

Solar loops *Gas can form arching loops that rise from areas of strong magnetism. The ends of the loops are attached to the Sun's magnetic poles.*

Spicules *Like a forest of small flames in constant motion, spicules actually rise several thousand kilometres from the solar surface.*

At the surface

The Sun has a surface but it is not solid like Earth's. It is made of gas bubbling and churning at 5,500°C (10,000°F)—hot enough to vaporise any solid. Solar storms erupt with bursts of radiation, while dark sunspots grow and then shrink over days and weeks, making the Sun a place of never-ending turmoil.

Granules *The Sun's surface is not smooth but is divided into cells 950 kilometres (600 mi) wide called granules, which are made of rising and sinking gas.*

SOLAR WINDS

The Sun's atmosphere, called the solar corona, is made of thin but hot gas. Gas streams from the corona in a constant wind of magnetically charged particles, which blows in all directions. Large solar flares on the Sun produce gusts in the solar wind that make it blow stronger.

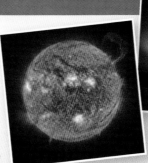

Mass ejections
Flares blow gas bubbles from the Sun. The charged particle clouds shoot across the Solar System in "coronal mass ejections".

Magnetic storm
If a solar "mass ejection" cloud reaches Earth it can disturb Earth's magnetic field and disrupt radio communication. Around the poles, it triggers displays of auroras.

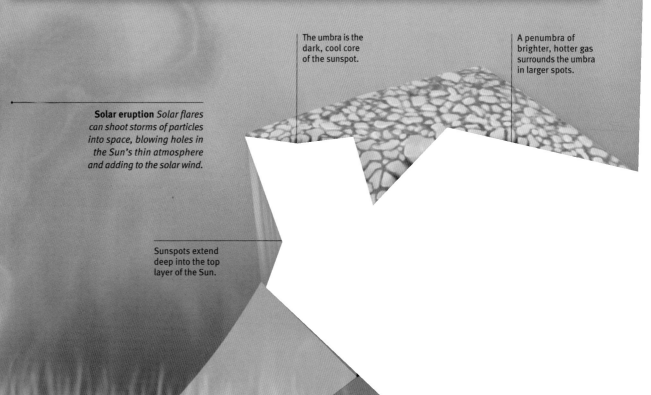

The umbra is the dark, cool core of the sunspot.

A penumbra of brighter, hotter gas surrounds the umbra in larger spots.

Solar eruption *Solar flares can shoot storms of particles into space, blowing holes in the Sun's thin atmosphere and adding to the solar wind.*

Sunspots extend deep into the top layer of the Sun.

Solar flares *In some areas, the Sun's surface erupts in brilliant white flares that release explosions of intense radiation in a few minutes.*

Sunspots *These dark regions occur where surface gases stop boiling up and down. With no supply of hot gas coming from below, the surface cools off.*

Sun myths

Ancient civilisations often considered the Sun as a powerful god, the giver of light and life. In Greece, people worshipped Apollo, god of the Sun.

Apollo

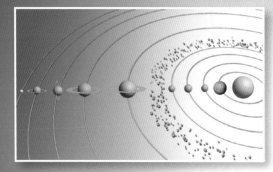

MERCURY: THE FACTS

ORIGIN OF NAME: Messenger of the Roman gods

DISCOVERED: Known since antiquity

DISTANCE FROM THE SUN: 58 million km (36 million mi)

MASS: 0.05 (Earth = 1)

VOLUME: 0.06 (Earth = 1)

DIAMETER: 4,879 km (3,032 mi)

SURFACE TEMPERATURE: −180°C to 430°C (−290°F to 800°F)

NUMBER OF MOONS: 0

LENGTH OF DAY/YEAR: 58.6 Earth days/87.6 Earth days

The Closest In
Mercury

The closest planet to the Sun is also the smallest of the major planets. Contrary to what astronomers once thought, Mercury does turn slowly on its axis (once every 58 Earth days), exposing all places to the scorching Sun by day and cold darkness by night. No other planet has such an extreme range of temperatures from night to day. Being closest to the Sun, Mercury also travels the fastest, taking just 88 Earth days to whip once around the Sun. These two factors make Mercury a hard planet to find in the sky—it appears for only a few weeks each year low in our evening or morning sky.

Caloris impact
A huge impact early in Mercury's history helped shape the thin crust that surrounds Mercury's giant iron core.

Contrasting worlds

The two inner planets are very different. Mercury is an airless, cratered world where, like our Moon, little has changed in billions of years. Venus has a thick, cloudy atmosphere and giant volcanoes that spew molten lava, reshaping the planet's hot surface and erasing any craters that do form.

A colliding asteroid blasts out the Caloris impact basin.

Direction of shockwaves

Shockwaves

Crust

Mantle

Core

Shockwaves crack the crust.

Jumbled chaotic terrain forms opposite the impact.

Highlands *As on our Moon, Mercury's highlands are generally older and more heavily cratered than the low plains.*

Rupes *These cliffs are 1–4 kilometres (0.5–2.5 mi) high. They snake across the planet and may have formed when Mercury cooled and shrank, cracking the crust.*

Impact craters *Impacts long ago pockmarked much of Mercury's surface. With no air or water to erode them, craters remain for billions of years.*

Volcanic plains *When large impacts smashed through the crust early in Mercury's history, lava erupted and poured across the surface, forming smooth plains.*

GASES · CRUST · MANTLE · OUTER CORE · INNER CORE

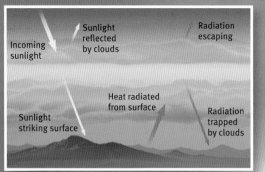

Incoming sunlight · Sunlight reflected by clouds · Radiation escaping · Heat radiated from surface · Radiation trapped by clouds · Sunlight striking surface

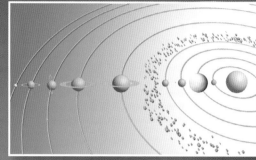

VENUS: THE FACTS

ORIGIN OF NAME:	Roman goddess of love and beauty
DISCOVERED:	Known since antiquity
DISTANCE FROM THE SUN:	108 million km (67 million mi)
MASS:	0.82 (Earth = 1)
VOLUME:	0.86 (Earth = 1)
DIAMETER:	12,104 km (7,521 mi)
SURFACE TEMPERATURE:	460°C (870°F)
NUMBER OF MOONS:	0
LENGTH OF DAY/YEAR:	243.0 Earth days/224.7 Earth days

Greenhouse planet

Venus is hotter than Mercury because Venus has an atmosphere that traps heat, like a greenhouse. With its thick atmosphere, Venus has much stronger greenhouse warming than Earth.

Lightning *High in the dense atmosphere, lightning may flash out of the yellow, sulphuric acid clouds. However, the surface itself is dry—too hot for even sulphuric acid to rain like water does on Earth.*

Volcanoes *These mountains can tower several kilometres above the plains. Lava once poured from them and flowed across the surface like rivers. Perhaps it still does.*

Volcanic domes *These low hills, several kilometres wide, are created where sticky lava pushes up the surface, oozes out slowly and cools.*

Sulphur deposits *Sulphur—which exists as a solid on Earth—may exist in liquid form on the superheated surface of Venus.*

Volcanic plains *Venus has a much younger surface than our Moon or Mercury, thanks to a relatively recent flow of lava across the planet.*

The Evening Star
Venus

In our night sky, Venus looks like a beautiful and bright evening or morning star. Though named for the goddess of beauty, the planet is anything but pretty when we look at it up close. Any robot probe that lands is fried by oven-like temperatures on the surface and crushed by an atmosphere 90 times denser than Earth's. The atmosphere is made of choking carbon dioxide, laced with clouds of burning sulphuric acid. Venus is only very slightly smaller than Earth, but because it is closer to the Sun, a year on Venus is much shorter than an Earth year, just 224 Earth days long. However, Venus rotates once every 243 Earth days, and backwards.

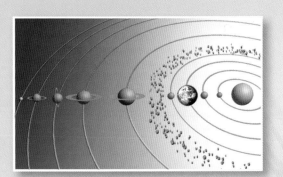

EARTH: THE FACTS

ORIGIN OF NAME: Ancient Saxon word for "ground" or "earth"

DISCOVERED: Known since antiquity

DISTANCE FROM THE SUN: 149.6 million km (92.9 million mi)

VOLUME: 1,086 billion cubic km (260 billion cu mi)

GRAVITY: 9.8 m/sec² (32.15 ft/sec²)

DIAMETER: 12,756 km (7,926 mi)

SURFACE TEMPERATURE: −88°C to +58°C (−126°F to +136°F)

NUMBER OF MOONS: 1

LENGTH OF DAY/YEAR: 24 hours/365.25 days

Earth and Its Moon

Our planet is the largest of the four rocky planets (Mercury, Venus, Earth and Mars), and is unique amongst those worlds in having a single large moon. Compared to the size of Earth, the Moon is so big that astronomers sometimes think of Earth and the Moon as a "double planet". Both worlds are made of silicate rocks, from the surface crust down to the bottom of a deep mantle layer. However, Earth may have a hot, molten nickel-iron core, unlike the Moon. Earth also has an atmosphere, weather and oceans so extensive that water covers 70 per cent of the planet's surface.

How the Moon formed

For centuries, it has been a mystery how Earth obtained its large satellite. Evidence from Moon rocks brought back to Earth by astronauts in the 1970s suggests that shortly after Earth formed it was hit by a rogue proto-planet. The Moon formed from the debris created by this titanic collision.

Impact *A proto-planet smashes into the early Earth, perhaps within the first 100 million years after Earth formed about 4.6 billion years ago.*

The Sun and the seasons

Seasons occur because Earth is tipped over on its axis. When the northern half of Earth tips towards the Sun, the Northern Hemisphere has warm summer days, but the Southern Hemisphere has cold winter days. Half a year later, the seasons are reversed.

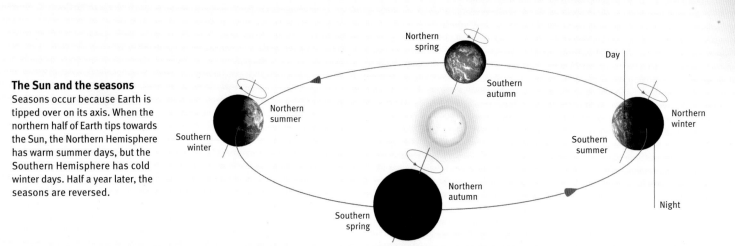

Northern spring

Southern autumn

Day

Northern summer

Southern winter

Northern winter

Southern summer

Northern autumn

Night

Southern spring

The Moon and the tides

The Moon's gravity tugs at Earth, raising the oceans in two bulges on opposite sides of the planet. As Earth turns, the sea level rises and falls twice a day, creating high and low tides along shorelines.

SOLAR ECLIPSES

When the Moon comes between Earth and the Sun, it creates a solar eclipse. People within the outer part of the Moon's shadow see part of the Sun covered; people in the narrow inner shadow see all of the Sun covered.

Partial eclipse | Total eclipse

Earth

Sun

Moon

Debris *A ring of debris forms around the primitive, still-molten Earth, giving the planet a short-lived ring system like Saturn's.*

Today *The Moon forms out of the fragments of the impact—debris from both the shattered proto-planet and Earth's top layers.*

New Moon	Waxing Crescent	First Quarter	Waxing Gibbous	Full Moon	Waning Gibbous	Last Quarter	Waning Crescent	New Moon

Phases of the Moon

As the Moon revolves around us, we see different amounts of the Moon lit by the Sun. When the Moon lies between us and the Sun, the side of the Moon facing us is dark—the New Moon. When it is opposite the Sun, the Moon appears fully lit—the Full Moon.

GASES

CRUST

MANTLE

CORE

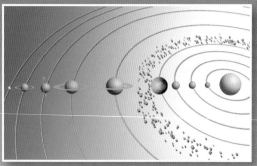

MARS: THE FACTS

ORIGIN OF NAME:	Named after the Roman god of war
DISCOVERED:	Known since antiquity
DISTANCE FROM THE SUN:	228 million km (142 million mi)
VOLUME:	0.15 (Earth = 1)
GRAVITY:	0.38 (Earth = 1)
DIAMETER:	6,792 km (4,220 mi)
SURFACE TEMPERATURE:	−125°C to +24°C (−195°F to +75°F)
NUMBER OF MOONS:	2
LENGTH OF DAY/YEAR:	24.6 hours/687 Earth days

Mars the Red Planet

Mars, the fourth planet from the Sun, is most like Earth. It is a rocky world with an iron core. It has an atmosphere, weather, seasons, polar ice caps and abundant frozen water. A Martian day, or sol, is just 41 minutes longer than an Earth day, but people could not live there without spacesuits. It is a bitterly cold desert of rusty rock and red sand—hence its nickname, the Red Planet. The air is poisonous carbon dioxide gas. A typical day is colder than the Antarctic in winter, with wild dust storms. Despite the harsh conditions, missions may land astronauts on Mars in 20 to 30 years.

Navcam *Black-and-white cameras take stereo 3-D images of nearby terrain. The rover navigates with the help of these images.*

Pancam *Like a pair of eyes on top of a ship's mast, colour cameras swivel 360 degrees to take panoramic images of the landscape.*

Antennae *Via its high- and low-power antennae, the rover sends data back to Earth and to probes orbiting above Mars.*

Solar panels *These provide electrical power to run the rover during the day and recharge its batteries to function at night.*

Electronics box *Computers and electronics are in an insulated, heated box. This protects them from the cold air.*

Hazcam *These cameras look out for boulders and other hazards too challenging for the rover.*

Six-wheel drive *Electric motors on each of the six wheels power the rover over small rocks and sand dunes.*

Roving on Mars

In January 2004, two golfcart-size robotic rovers landed on opposite sides of Mars. Remote-controlled from Earth, Spirit and Opportunity started exploring the Martian world of craters, sandy plains, hills, canyons and extinct volcanoes. Both rovers were still operating two years later, having withstood the harsh conditions, travelled many kilometres from their landing sites, and collected data and images of rocks and soils. The rovers succeeded in finding evidence that water once flowed across cold, dry Mars.

WATER ON MARS

Mars is too cold for liquid water, but orbiting space probes have found polar ice caps, icy fogs and ice patches. More is buried as permafrost. If, billions of years ago, Mars was warmer and its water formed lakes and seas, then life—which needs liquid water to survive—may have started. Did it die out as Mars froze? The answer remains a mystery.

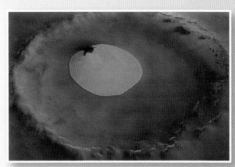

The Mars Express orbiter discovered a field of ice at the bottom of this far northern crater.

Icy fog shrouds Valles Marineris, a canyon so long it would stretch right across the USA.

Dust devils *Warming air rises, picks up dust and creates whirling dust devils that race across the landscape.*

Movable arm *A robotic arm can bend and reach so that instruments on the end can touch nearby rocks and soil.*

Instruments *A microscope camera takes close-ups of rocks, while a grinding tool drills into them.*

Rocks in the landscape

Both Spirit and Opportunity found rocks formed by volcanic eruptions. But Opportunity also quickly discovered rocks covered in "blueberries". These small, round nodules are made of a mineral that forms in water. The "blueberries", and other tell-tale signs, such as layers in the rocks and on crater walls, suggest that water once flowed on Mars, perhaps as salty lakes or seas. Mars might once have been warmer, wetter and more like Earth.

GASES

LIQUID HYDROGEN

METALLIC HYDROGEN

CORE

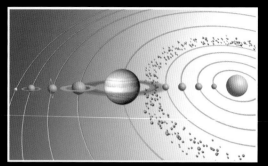

JUPITER: THE FACTS

ORIGIN OF NAME:	King of Roman gods
DISCOVERED:	Known since antiquity
DISTANCE FROM THE SUN:	779 million km (484 million mi)
VOLUME:	1,321 (Earth = 1)
GRAVITY AT CLOUDTOPS:	2.5 (Earth = 1)
DIAMETER:	142,980 km (88,844 mi)
CLOUDTOP TEMPERATURE:	−110°C (−160°F)
NUMBER OF MOONS:	63
LENGTH OF DAY/YEAR:	9.9 hours/11.86 Earth years

Jupiter and Io

If you were standing on Io, Jupiter's innermost large moon, the gas planet would loom large in the sky. Its bright and dark cloud belts would provide endless fascination as they swirl and twist into storms. From there you would also be able to see Jupiter's faint single ring made of fine dust particles.

A Giant of Gas

Jupiter

The Solar System's giant is Jupiter, a planet so large it could contain all the planets combined. Being so big, Jupiter has the strongest gravity of all the planets and thus holds on to the most moons. Jupiter spins so rapidly on its axis that it bulges out at the equator; a Jovian day lasts less than 10 hours. Its fast rotation also whips its clouds into broad planet-circling belts, dotted with immense storms. These clouds are only the top layers of a planet made almost entirely of hydrogen gas—there is no solid surface that anyone could land on. In many ways, Jupiter is more like a small failed star than a planet like Earth.

Jupiter's moons

Jupiter's four large "Galilean" moons are named for Galileo, the astronomer who discovered them. All are larger than our Moon, and Ganymede is the biggest moon in the Solar System.

Ganymede

Callisto

Io

Europa

Clouded realm

Jupiter's colourful clouds are made of mixtures of ammonia, sulphur and water vapour. Temperatures at the tops of the clouds are frigid, but increase as you descend deeper into the atmosphere.

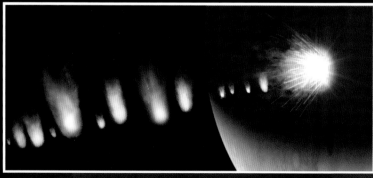

Collision course

In July 1994, the many pieces of an icy comet called Shoemaker-Levy 9 slammed into Jupiter one after the other. They caused a series of explosions that left dark spots in Jupiter's clouds.

Icy Europa

A crust of cracked ice that might be several kilometres thick covers a global ocean of liquid water on this strange moon.

Size of the Great Red Spot

Spinning around like a giant hurricane, the Great Red Spot has been raging for several centuries. It is so large it could swallow two Earths. It spins anticlockwise every seven days and changes colour from deep red to pale pink over many years. But what makes it this colour and keeps it spinning remains a mystery.

Giant storm The Great Red Spot is 24,000 km (15,000 mi) long; Earth is 12,756 km (7,926 mi) in diameter.

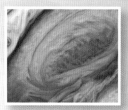

Great Red Spot

Earth

In false colour The colour looks unnatural, but colour filters reveal more detail than our eyes can see.

Cloud belts *When we look at Jupiter, all we see are the tops of bright and dark cloud bands that circle the planet.*

Bright clouds *The white zones are high, cold clouds made of ammonia crystals. Deep down are clouds of water vapour like on Earth.*

Great Red Spot *Strong winds churn up many swirling storms in Jupiter's atmosphere. The biggest is the Great Red Spot.*

Volcanoes *These spew plumes of molten sulphur into the sky and across the surface, colouring Io in hues of yellow and red.*

Standing on Io *Superhot volcanoes constantly erupt on Io. The moon is thought to be the most volcanically active world in our Solar System.*

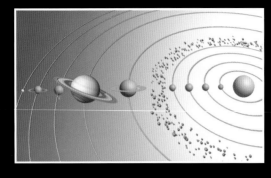

ORIGIN OF NAME:	Roman god of agriculture
DISCOVERED:	Known since antiquity
DISTANCE FROM THE SUN:	1,427 million km (887 million mi)
VOLUME:	752 (Earth = 1)
GRAVITY AT CLOUDTOPS:	0.9 (Earth = 1)
DIAMETER:	120,540 km (74,900 mi)
CLOUDTOP TEMPERATURE:	−140°C (−220°F)
NUMBER OF MOONS:	56
LENGTH OF DAY/YEAR:	10.6 hours/29.5 Earth years

Bobbing in water
Saturn is made mostly of light hydrogen gas. If you could find a container big enough to hold Saturn, it would float in water.

Lord of the Rings

Saturn

While other planets have rings, Saturn's are the brightest and most magnificent. The rings are three-quarters of the distance from Earth to the Moon in width, and yet are no more than a city block thick. Viewed from Earth, the main system seems divided into three prominent rings. Close-up images from space probes reveal that each ring is really thousands of strands of orbiting ice particles. Saturn is a gas giant similar to Jupiter, made mostly of hydrogen gas. It is topped with cloud belts and swirling storms of ammonia and water ice crystals. Despite its huge size Saturn spins rapidly; each day is only 10.6 hours long.

F Ring *Two little "shepherd moons" orbiting either side of the F ring keep its particles confined to a thin thread.*

Ring gaps *The gravity of small moons sweeps some areas free of particles, creating gaps and divisions in the rings.*

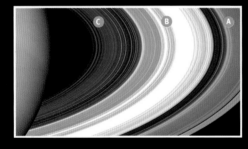
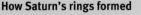

Alphabet rings
Saturn has three main rings: the outer A ring, contained by a small moon orbiting outside; the dense and bright middle B ring, made of particles packed closely together; and a less dense and darker inner C ring, partly transparent because it has fewer ice particles in it.

How Saturn's rings formed

A comet or asteroid collides with an icy moon, perhaps within the last few hundred million years.

The impact smashes both into billions of icy pieces that could not re-form into another moon.

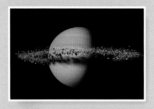

The icy debris spreads around the planet and more collisions grind up the ice particles further.

The gravity of other moons shapes the rings, while future collisions may add to the rings.

Plasma wave antennae *Three whiplike antennae detect radio waves from the Sun and Saturn.*

Magnetometer *This is an instrument on the end of a long boom that measures Saturn's magnetic field.*

High-gain antenna *This antenna radios images and data back to Earth.*

Huygens probe *This ride-along probe was released in 2004 to land on Saturn's moon Titan.*

Shooting over the rings

After a seven-year journey, the Cassini spacecraft fired its engines to enter orbit around Saturn on 1 July 2004. With engines blazing, Cassini skimmed just above the rings. As it orbits Saturn many times over several more years, the probe will explore Saturn's complex rings and icy moons.

Engines *A large rocket engine placed Cassini into orbit, while smaller rockets turn the probe towards targets.*

Ring particles *These consist of billions of chunks of ice, ranging from snowflakes to giant boulders. Each orbits Saturn like little moons.*

Cassini's Division *Named for its seventeenth century French discoverer, Jean Cassini, this ring gap is as wide as North America.*

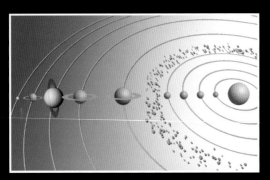

Icy twins

Similar in size and composition, Uranus and Neptune are twin "ice giant" planets. Unlike Jupiter and Saturn, which are mostly hydrogen gas, Uranus and Neptune seem to be made mostly of a slushy ice mixture of frozen water, methane and ammonia.

URANUS: THE FACTS

ORIGIN OF NAME: Greek god of the sky

DISCOVERED: 1781 by William Herschel

DISTANCE FROM THE SUN: 2,871 million km (1,784 million mi)

VOLUME: 63.1 (Earth = 1)

GRAVITY: 0.90 (Earth = 1)

DIAMETER: 51,120 km (31,764 mi)

CLOUDTOP TEMPERATURE: −215°C (−350°F)

NUMBER OF MOONS: 27

LENGTH OF DAY/YEAR: 17.2 hours/84 Earth years

Clouds and atmosphere
The atmosphere lacks clouds or storms, probably because Uranus has no internal source of heat.

Rings *At least 11 rings circle Uranus. They are also tipped over to align with the planet's equator.*

The Sideways Planet

Uranus

The third largest planet in the Solar System, Uranus is an odd world that rotates around an axis that is flipped sideways. At times during the Uranian year, one pole points towards the Sun, so Uranus "rolls" around it. Uranus also has the third-greatest number of moons, with at least 27 orbiting satellites. The rings of Uranus are a series of thin, threadlike strands made of dark, icy particles. Compared with the swirling clouds of Jupiter and Saturn, Uranus's atmosphere appears relatively nondescript and featureless, with only a few isolated white clouds ever seen.

Moons in the rings
Shepherd moons orbiting on either side of each ring may keep them so narrow.

Inside the planets

Rocky planets, gas giants and ice giants differ not only in size but also in internal composition. Earth's crust and mantle are rocks and minerals. Jupiter is mostly hydrogen while Neptune is mainly ice.

HOW URANUS GOT ITS TILT

Early in its history, Uranus may have been hit by a proto-planet that knocked the planet over on its side, where it remains today. Its axis is tipped 97.9 degrees, compared to Earth's tilt of 23.5 degrees.

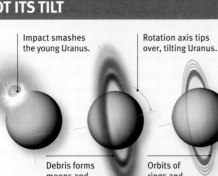

Impact smashes the young Uranus.

Rotation axis tips over, tilting Uranus.

Debris forms moons and rings.

Orbits of rings and moons also tilt.

Earth has a molten outer core of iron and nickel. The inner core is solid.

Jupiter has a rocky centre inside a mantle of metallic hydrogen.

Neptune has a rocky centre inside a mantle of hard ice.

GEYSERS ON TRITON

Despite its supercold surface (−235°C/−391°F), Neptune's moon Triton has enormous geysers that erupt with plumes of liquid nitrogen, which rains down to coat the surface with nitrogen frost.

NEPTUNE: THE FACTS

ORIGIN OF NAME: Roman god of the sea
DISCOVERED: 1846 by Johann Galle
DISTANCE FROM THE SUN: 4,497 million km (2,794 million mi)
VOLUME: 57.7 (Earth = 1)
GRAVITY: 1.14 (Earth = 1)
DIAMETER: 49,530 km (30,776 mi)
CLOUDTOP TEMPERATURE: −195°C (−320°F)
NUMBER OF MOONS: 13
LENGTH OF DAY/YEAR: 16.1 hours/165 Earth years

The Smallest Giant
Neptune

Like Uranus, Neptune is made mostly of ice, topped by a layer of gas, and probably with a small rocky core. Its system of thin rings are so dark they are difficult to see from Earth. Yet Neptune is different because it emits heat from its interior. This warmth rising through the atmosphere stirs up Neptune's clouds and creates weather and storms. Number four in size, Neptune is also fourth in number of moons, hosting a family of at least 13 satellites. Only one, Triton, is a large moon, with a diameter about three-quarters the size of our own Moon.

Weighing up the moons
Even though Uranus has twice as many moons as Neptune, Neptune's moons weigh more because of Triton's large size and mass. Both their moons are made mostly of ice.

Uranus: 27 moons   Neptune: 13 moons

Rings At least five rings orbit around Neptune. They were discovered in 1989 by the Voyager 2 probe.

Moons in the rings Four shepherd moons have been found orbiting within Neptune's rings.

Clouds and atmosphere The atmosphere often has white clouds and occasionally hurricane-like dark spots.

Gusting winds Winds on Neptune are the fastest in the Solar System, at about 2,400 kilometres per hour (1,450 mph).

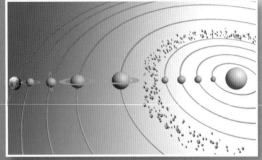

PLUTO: THE FACTS

ORIGIN OF NAME: Roman god of the underworld

DISCOVERED: 1930 by Clyde Tombaugh

DISTANCE FROM THE SUN: 5.9 billion km (3.7 billion mi)

VOLUME: 0.006 (Earth = 1)

GRAVITY: 0.06 (Earth = 1)

DIAMETER: 2,390 km (1,485 mi)

SURFACE TEMPERATURE: −230°C (−380°F)

NUMBER OF MOONS: 3

LENGTH OF DAY/YEAR: 6.39 Earth days/248 Earth years

OBJECTS IN ORBIT

Like comets, Pluto, Eris and many Kuiper Belt objects go around the Sun in long elliptical orbits, taking centuries to make one circuit of the Sun. By comparison, the orbits of the eight major planets are more like circles.

Eris

Pluto

Neptune

Halley's Comet

Kuiper Belt

Pluto and Beyond

...red the ninth planet when it was discovered in 1930,
...ow called a dwarf planet. It is just one member of
...tion of small, icy worlds at the edge of the Solar
...s called the Kuiper Belt. Pluto is tiny—at 2,390
...5 mi) in diameter it is smaller than our Moon.
...d has three moons of its own: Charon is
...hile Nix and Hydra are only a few dozen
...other Kuiper Belt object known only as
...s, while Eris, larger than Pluto and also
...has one moon called Dysnomia.

Country-size dwarf
Pluto and its largest moon
Charon are so small that both
could fit across the width of
the USA, with room to spare.

Sedna

Floating in the Kuiper Belt
Thousands of small worlds made of ice orbit beyond
Neptune. Most are a few kilometres across, but some are
nearly as large as, if not larger than, Pluto. No spacecraft
have visited any of these worlds, so the images here
depict what they might look like up close.

Sedna *Sedna is now three
times farther away than Pluto
and takes 10,500 years to
orbit the Sun. It is the coldest
place ever discovered in
our Solar System.*

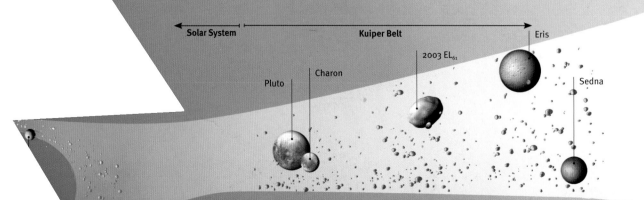

Solar System | Kuiper Belt | Eris

Pluto

Charon

2003 EL$_{61}$

Sedna

2003 EL₆₁

Unnamed moon #2

Unnamed moon #1

2003 EL₆₁ system *Still awaiting an official name, 2003 EL₆₁ has two satellites and spins so fast (every four hours) that it is elongated into a potato shape.*

Eris

Eris system *Discovered in 2005, Eris is slightly larger than Pluto. Rather than call Eris the 10th planet, Eris and Pluto are now both called dwarf planets.*

Dysnomia

Pluto system *Pluto has a thin atmosphere and it likely has frost or snow on its surface. Its three moons may have formed from the debris left over from when Pluto was hit by another Kuiper Belt object billions of years ago.*

Pluto

Charon

Nix

Hydra

Photo-shy Pluto

The best pictures of Pluto, from Hubble, show only bright and dark markings. We'll see more in 2015 when the New Horizons probe flies past Pluto and Charon.

EAGLE NEBULA: THE FACTS

ORIGIN OF NAME: Named because it resembled a flying eagle

DISCOVERED: 1794 by Charles Messier

DISTANCE FROM EARTH: 6,500 light-years

CONSTELLATION: Serpens Cauda, the Serpent Tail

LOCATION: Sagittarius Arm of the Milky Way

TYPE: Emission nebula with star cluster

DIAMETER: 70 light-years

BEST SEEN: May to August

PHOTO: Young stars light up gas and dust in the nebula

Glowing tops *The tops of the columns are lit by the ultraviolet light from nearby stars.*

Pillars of Creation

Eagle Nebula

Stars are born deep inside immense regions of gas and dust like the Eagle Nebula. The Eagle can be seen through a backyard telescope, but it took the power of the orbiting Hubble Space Telescope to reveal how the stars develop. Radiation and winds, blowing from nearby hot blue stars just 5 million years old, burn away the gas and dust in the nebula. This leaves columns of dark, denser gas that resist erosion from the young stars. New stars form in these protective cocoons. The blast wave from a nearby exploding supernova may topple the Eagle's pillars of gas in another thousand years, exposing newborn stars forming within the nebula.

Shaped like an eagle
Astronomers imagine this nebula looks like an eagle in flight. Can you see it? In this wide-angle photo taken from Earth, the "pillars of creation" are the small angled fingers in the center.

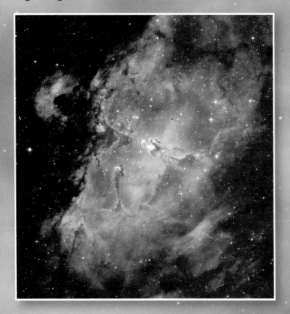

Hot star *Radiation from this hot blue star is hollowing out the side of the largest dust column.*

High tower *The tallest pillar towers four light-years high. This is the same distance as from the Sun to Proxima Centauri, its nearest star.*

Pillars of creation

A photograph from the Hubble Space Telescope revealed amazing details in the Eagle Nebula's central columns of cool gas and dust. At the tops of these "pillars of creation", small fingers of gas are breaking away to form free-floating blobs called Evaporating Gaseous Globules, or EGGs.

Star birth

Inside the Eagle Nebula we see dense masses of gas (EGGs) breaking off the main pillars. Stars then grow inside these protective EGGs. Perhaps our own Sun and planets formed in a similar process.

Stellar radiation erodes fingers of gas.

EGG breaks away and shrinks to form star.

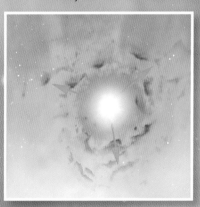

New star blows away remaining gas.

Streaming gas *Hot gas, warmed by nearby stars, streams away from the column tips.*

Enormous EGGs *While each EGG appears to be tiny, it is about the size of our Solar System and houses a new star.*

Fingers and EGGs *Dense fingers of gas and bloblike EGGs protrude from the tips of the columns.*

Wearing down *The dark columns erode away, leaving enough EGGs to form dozens of new stars.*

Blown away *Sections of the pillars are blown into space before they develop into new stars.*

MILKY WAY: THE FACTS

ORIGIN OF NAME:	From Greek and Latin words for "milk"
DISCOVERED:	Known since antiquity
NUMBER OF STARS:	200 to 400 billion
TYPE:	Type SBbc barred spiral galaxy
DIAMETER:	100,000 light-years for visible disk
DISTANCE OF SOLAR SYSTEM:	26,000 light-years to galaxy centre
LOCATION OF SOLAR SYSTEM:	Orion spur of Sagittarius spiral arm
BEST SEEN:	Summer, autumn and winter
PHOTO:	Diagram shows location of the Sun in the Milky Way

Our Galaxy the
Milky Way

Of all the billions of galaxies in the universe, the Milky Way is special because it is the one we live in. Galaxies are huge collections of stars, gas and dust that take different forms. The Milky Way is a barred spiral galaxy. It contains about 200 to 400 billion stars, each one orbiting around the galaxy's core. The Sun is just one of those stars. It is not at the centre of our galaxy, but lies in an outer spiral arm of the Milky Way. Astronomers think that a massive black hole sits at the galaxy's centre, devouring any star that comes too close to the powerful pull of its gravity.

Milky Way *Because we live within it, our galaxy appears now (as it will in the future) as a misty band of faint stars stretching across the sky.*

Tidal tails *Gravity might pull streams of stars away from both galaxies, creating new milky streams of stars across the night sky.*

PATHWAY IN THE SKY

From Earth, the Milky Way's spiral arms appear as a misty band of light stretched across the night sky. Many ancient civilisations included it in their myths and legends. To the Norse, it was the road to Valhalla, home of slain warriors. In China and Japan, it was the "river of heaven" or the "silver river". The ancient Greeks thought it was milk spilt by the goddess Hera—hence its name, the Milky Way.

Milky band At certain times of the year, the Milky Way stretches across the sky.

A slow dance

① Over three billion years, the two galaxies draw closer together.

② They begin to dance around each other.

③ Gravity pulls streams of stars out of each galaxy.

④ As long tails of stars are lost to space, the cores fall together.

⑤ After about a billion years, they merge to form one large, elliptical galaxy.

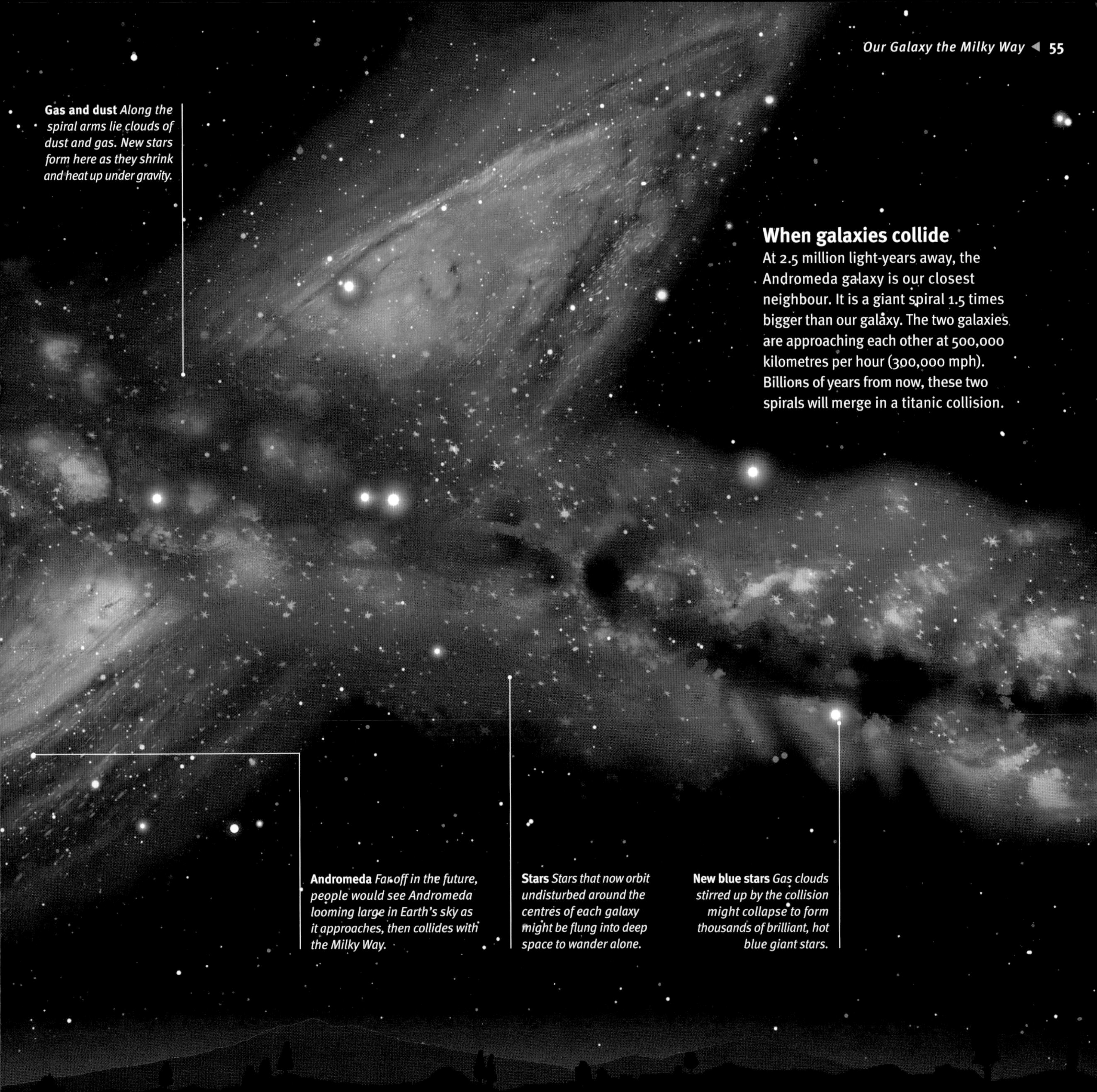

Gas and dust *Along the spiral arms lie clouds of dust and gas. New stars form here as they shrink and heat up under gravity.*

When galaxies collide
At 2.5 million light-years away, the Andromeda galaxy is our closest neighbour. It is a giant spiral 1.5 times bigger than our galaxy. The two galaxies are approaching each other at 500,000 kilometres per hour (300,000 mph). Billions of years from now, these two spirals will merge in a titanic collision.

Andromeda *Far off in the future, people would see Andromeda looming large in Earth's sky as it approaches, then collides with the Milky Way.*

Stars *Stars that now orbit undisturbed around the centres of each galaxy might be flung into deep space to wander alone.*

New blue stars *Gas clouds stirred up by the collision might collapse to form thousands of brilliant, hot blue giant stars.*

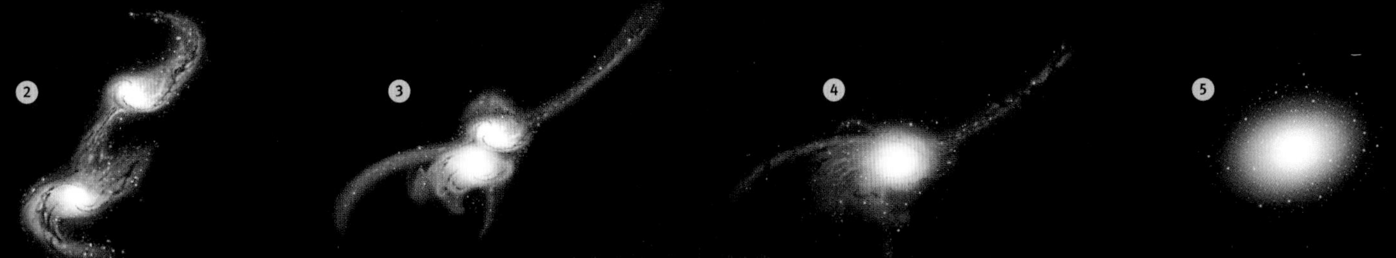

2

3

4

5

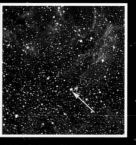

SUPERNOVA 1987A: THE FACTS

ORIGIN OF NAME: First supernova discovered in 1987

DISCOVERED: 23 February 1987, by Ian Shelton

DISTANCE FROM EARTH: 169,000 light-years

CONSTELLATION: Dorado, the Swordfish

LOCATION: Large Magellanic Cloud

TYPE: Type IIP Supernova

DIAMETER: Central ring 1 light-year across

BEST SEEN: December to March (from south of the Equator)

PHOTO: SN 1987a peeks from behind clouds of gas

Before and after
Before the explosion the unstable star shone dimly in a satellite galaxy of the Milky Way. It then blazed to become the brightest supernova seen in 400 years.

Anatomy of
Supernova 1987A

Big stars blow up at the ends of their lives. In 1987 the world watched as a giant star 169,000 light-years away exploded as a supernova. For a few weeks it shone with the energy of 100 million Suns. Even at its great distance, it was so bright that the supernova could be seen as a brilliant temporary star shining in the southern sky. Astronomers in Chile spotted the star on 23 February; it reached its peak intensity on 20 May. Since then, the Hubble Space Telescope has watched the supernova's remains expanding away from the explosion site. From the evidence, it appears that a giant blue star 20 times the mass of our Sun blew up, then collapsed into a neutron star.

Shattered remains

Close-up images from Hubble reveal the supernova surrounded by rings of material lit by blast waves from the explosion. The star's remnant core most likely shrunk under gravity to form a neutron star, a bizarre object the size of a city but containing as much matter as the Sun.

Building up to a blowup
The illustrations below show the lead up to a typical Type II supernova. 1987A falls into this category, but it was initially a blue giant.

Unstable star
A giant star burns so hot that in just a few million years it runs out of hydrogen and other fuel. The outpouring of energy that had kept the star inflated suddenly stops.

Core collapse
Without energy to keep it alive, gravity takes over, collapsing the star's core to a tiny size. It may shrink to a rapidly spinning compact star made of dense neutrons.

Bang!
Radiation from the collapsing core blasts the star's bloated outer layers into space. The supernova's expanding debris briefly shines brighter than 100 million Suns.

1. **Neutron star** *Though it has not yet been detected, astronomers think a neutron star now lies hidden deep inside the supernova debris.*

2. **Debris cloud** *The dying blue star blew off its outer layers, creating a cloud of debris expanding away from the collapsed star.*

3. **Central ring** *Blast waves from the supernova are now smashing into a ring of material blown off the star 20,000 years before the explosion, heating the ring's gas.*

4. **Hot spots** *Looking like bulbs on a string of Christmas lights, bright spots show where the colliding blast wave lights up the gas in the central ring.*

5. **Outer rings** *Two mysterious rings, one in front and one behind the supernova, may be material blown from the doomed star's poles or from a destroyed companion star.*

6. **Other stars** *These bright stars look like they belong to the supernova debris, but they are really ordinary stars in the foreground.*

Cataclysmic binaries

Some stars become supernovas because they are part of a two-star system. A dwarf star robs material from a giant companion, becoming overweight and unstable. The dwarf then explodes, destroying itself and freeing its companion to fly off into space.

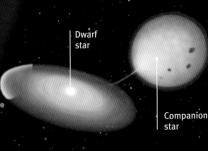

Dwarf star

Companion star

A white dwarf in a binary system steals gas from its larger companion, a red giant.

The bloated dwarf blows up, tearing apart the system and flinging the giant into space.

CYGNUS X-1: THE FACTS

ORIGIN OF NAME: First X-ray source found in Cygnus

DISCOVERED: 1972 by Tom Bolton

DISTANCE FROM EARTH: 8,200 light-years

CONSTELLATION: Cygnus, the Swan

LOCATION: Sagittarius Arm of the Milky Way

TYPE: Stellar-mass black hole

DIAMETER: 30 to 60 km (20 to 40 mi) for black hole

BEST SEEN: July to October

PHOTO: Seen in X-rays, a disk of gas surrounds Cygnus X-1

Into a Black Hole

Cygnus X-1

When the biggest stars explode, the leftover core collapses into an object so small and dense its gravity becomes too strong for even the fastest thing in the universe—light—to escape the object's grasp. This is a black hole. Anything coming too close to the black hole is drawn in and crushed out of existence. Cygnus X-1 was the first confirmed black hole to be discovered. The hole orbits a large, blue supergiant star, sucking material off the unfortunate star. Luckily, our Sun is much too small to ever turn into a black hole. Nor is it likely that Earth and the Solar System would ever bump into one—black holes are rare and, as best we know, none lurk close to us.

Distorting light
A black hole's intense gravity bends beams of light that pass near the hole. Acting like a magnifying lens, a black hole will distort images of distant objects behind it.

Anatomy of a black hole

Because black holes emit no light, they can only be detected if they orbit another star—a source of gas and dust that swirls down the hole. Telescopes that detect radiation from this material allow us to "see" black holes.

Companion blue star
The black hole in Cygnus X-1 orbits a blue supergiant star 25 to 30 times heavier and many times bigger than our Sun.

KINDS OF BLACK HOLES

Like a heavy object sitting on a rubber sheet, everything made of matter (people, planets, stars) warps space and time. The more massive the object, the bigger and deeper the space warp.

Normal-size star
Stars like the Sun create a shallow dip in the fabric of space and time.

Stellar-mass black hole
Collapsed stars like Cygnus X-1 make black holes that punch a small hole in space.

Supermassive black hole
Giant black holes in galaxy cores produce a deep, large puncture in space.

Fast orbit *The blue supergiant orbits the black hole every 5.6 days.*

Falling into a black hole

Earth could orbit a black hole beyond the event horizon. But get any closer, and the difference in gravity from one side of Earth to the other first stretches, then rips our doomed planet apart into subatomic particles.

At the event horizon | Stretched like spaghetti | Ripped apart

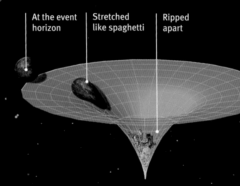

Infalling gas *Gas orbiting at the inner edge of the accretion disk (about 160 kilometres/100 mi above the hole) eventually falls into the hole.*

Black hole *The hole in Cygnus X-1 has a mass of 7 to 15 Suns. This is all that remains of a much larger star that exploded long ago.*

Polar jets *Jets of energised particles shoot away from the north and south poles of the Cygnus X-1 system.*

X-ray emission *As the gas swirls furiously down the hole, it heats up to 100 million degrees and emits intense X-ray radiation.*

Event horizon *Though it would not be visible as a boundary, the event horizon is the point of no return—nothing that passes this limit can ever escape the hole's gravity.*

Stream of gas *The black hole strips gas off the outer layers of the blue supergiant star, feeding the black hole.*

Accretion disk *Material pulled from the blue supergiant goes into orbit around the hole. It forms a flat disk of hot gas, like water down a drain.*

Our Amazing
Universe

THE MILKY WAY GALAXY

The view from above
Looking down on the Milky Way, we would see a giant rotating pinwheel. Spiral arms slowly turn clockwise around a bright centre packed with stars. The stars closest to the centre, where gravity is strongest, orbit faster than those out near the edge.

1. **From the side** Spiral arms of the Milky Way stretch out in a thin, flat disk from a bulging central core of stars.
2. **Globular clusters** About 200 globular clusters orbit the galaxy, buzzing around its core like bees round a hive.

3. **Galactic proportions** The Milky Way is so big that the fastest thing in the universe, a beam of light, takes 100,000 years to get from one side to the other.
4. **Spiral arms** These are bands of brilliant stars and nebulas. The gaps between the arms have dimmer stars and fewer nebulas.
5. **Inner core** This is packed with stars. If we could view it from near the core, the night sky would be thick with white, yellow and red lights.
6. **Black hole** A massive black hole lurks at our galaxy's centre. A swarm of stars and gas swirls around the hole, feeding it with new material.
7. **Solar System** The Sun is about halfway between the Milky Way's bright core and its dark, remote edge.

THEORIES OF THE UNIVERSE

Changing ideas
Throughout history, astronomers have created pictures, or theories, to help explain how large the universe is and what it looks like. These theories have changed many times.

Aristotle
In about 350 BC, Aristotle said that Earth must be the centre of the Solar System and the universe.

Copernicus
In 1543, Copernicus proposed that Earth orbited the Sun like the other planets.

Today
We now know our Sun is but one star in the Milky Way, itself one of many galaxies in the universe.

THE ANDROMEDA GALAXY

At Andromeda's core
Inside Andromeda's core lies a black hole as big as the orbit of Earth around the Sun. Around it race a ring of red stars outside a small swarm of faster-moving blue stars. These blue stars travel at 3.6 million kilometres per hour (2.2 million mph).

How fast they go
If Andromeda's blue stars were orbiting Earth, they would whip around us in just 40 seconds. The Moon takes a month to orbit.

KINDS OF STAR SYSTEMS

Stars in a system
Our Sun is a single star, but many other stars belong to systems. Two or more stars, often of very different size and colour, orbit each other.

Two stars dance around each other in a binary star system.

This triple star system orbits a common centre of gravity.

Two-star system
Some solar systems are very different from our own. This system has a giant, Jupiter-like planet that orbits two suns, not one.

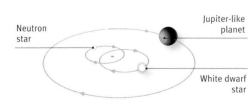

Neutron star

Jupiter-like planet

White dwarf star

MOVING STARS

The Big Dipper
Because all stars are slowly moving as they orbit the galaxy, the constellations will slowly change shape over thousands of years.

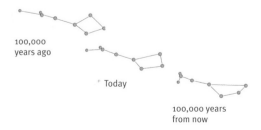

100,000 years ago

Today

100,000 years from now

COMPARING MOONS

Our Moon
3,475 km (2,160 mi)
in diameter

Ganymede
5,262 km (3,270 mi)
in diameter

The biggest moon
Ganymede is the Solar System's largest moon. It is bigger than Mercury and one-and-a-half times the size of the Moon.

TILTING PLANETS

Planets and their axes
Every planet rotates around an axis, but each planet's axis is tipped a different amount for reasons no one knows.

Mercury
Tilt of axis = 0°
No tilt to axis

Venus
Tilt of axis = 177.4°
Rotates backwards

Earth
Tilt of axis = 23.4°
Causes the seasons

Mars
Tilt of axis = 25.2°
Also has seasons

Jupiter
Tilt of axis = 3.1°
Almost no tilt

Saturn
Tilt of axis = 26.7°
Rings also tipped

Uranus
Tilt of axis = 97.9°
Rotates on its side

Neptune
Tilt of axis = 29.6°
Rings also tipped

HIGHS AND LOWS

Comparing Earth
The deepest canyon on Earth, the undersea Marianas Trench, is deeper than the biggest canyon on Venus. But the tallest mountain on Mars, the volcano Olympus Mons, towers above the highest peaks on Earth or Venus.

1. **Olympus Mons**
   Mars (26 km/16 mi)
2. **Maxwell Montes**
   Venus (10.8 km/6.7 mi)
3. **Mount Everest**
   Earth (8.8 km/5.5 mi)
4. **Sea level**
   0 km (0 mi)
5. **Diana Chasma**
   Venus (−2.9 km/−1.8 mi)
6. **Marianas Trench**
   Venus (−11 km/−7 mi)

TALLEST CLIFF

Verona Rupes
An ice cliff 20 kilometres (12 mi) high on Miranda, a moon of Uranus, is ten times the depth of the Grand Canyon.

BIGGEST CANYON

Valles Marineris
This Martian canyon is as wide as the USA. It is seven times wider and three times deeper than the Grand Canyon.

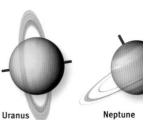

FIRST LOOK

First snapshot taken
The first photo of Earth from space was snapped by the TIROS weather satellite on 1 April 1960, from an altitude of 700 kilometres (435 mi).

VOYAGER

Farthest message
Two Voyager craft, now billions of kilometres away, are leaving the Solar System with a video disk and cover diagram to tell aliens about Earth.

LIFE ON OTHER WORLDS

The search for life
Life needs three main ingredients: liquid water, an energy source and the right chemicals to make cells. In our Solar System, four worlds, besides Earth, might qualify.

Best possibilities
Mars
Mars may have had a warm climate and liquid water in its ancient past.

Europa (Jupiter)
This moon may have an ocean warmed by internal heat.

Enceladus (Saturn)
Enceladus has geysers powered by an unknown heat source.

Titan (Saturn)
This moon has conditions similar to the primitive Earth.

Geysers on Enceladus

Titan's hazy atmosphere

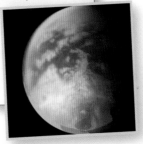

SPACE TOURISM

SpaceShipOne
In 2004, a rocket called SpaceShipOne became the first privately built craft to reach space. By 2009, a ship much like it will carry tourists on quick trips into space.

Tourism milestones
1991 Japanese journalist in Mir
2001 First space tourists: Dennis Tito (2001); Mark Shuttleworth (2002); Gregory Olsen (2005)
2004 SpaceShipOne first private craft in space
2009 First flights of Virgin Galactic planned

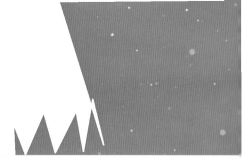

accelerating universe theory This suggests the universe is expanding in size at a faster and faster rate, propelled by a cosmic "antigravity" force.

accretion disk A disk of gas spinning rapidly around a powerful source of gravity, such as a black hole or a newly forming star.

asteroid A small rocky object, often irregular in shape, that orbits the Sun. Asteroids are too small to be called planets or dwarf planets.

asteroid belt Most asteroids orbit in a zone between the orbits of Mars and Jupiter, a belt containing hundreds of thousands of asteroids.

astronomy The science that studies the universe beyond Earth: other planets, stars and galaxies.

atmosphere The layer of gas that surrounds some rocky planets, such as Earth, or that forms the top layer of gas giant planets, such as Jupiter.

atmospheric pressure The measure of how much an atmosphere presses down onto the surface below; the deeper and denser the atmosphere, the greater the air pressure.

atom Each element of matter (such as hydrogen and helium) is made of atoms, with a nucleus made of protons and neutrons orbited by electrons. The number of protons an atom has determines what element it is.

aurora Auroras are glowing bands of light formed high in an atmosphere by particles from the Sun colliding with atoms in the atmosphere. On Earth, they are known as the northern or southern lights.

axis An imaginary line through the poles, around which a planet or star rotates.

Big Bang theory This proposes that, based on much evidence, all energy and matter in the universe today emerged from a single "explosion" of space and time about 13.7 billion years ago.

binary star A system of two stars orbiting around each other. Also known as a double star system.

black hole A region of matter so dense that its gravity is too strong for even light to escape.

carbon An element made of six protons and six electrons, usually a solid, that is essential for life as we know it.

carbon dioxide A molecule made of carbon and oxygen. It is the main gas in the atmospheres of Venus and Mars.

comet An icy object that can begin to vaporise when it gets close to the Sun, forming long gas and dust tails.

condensation The process by which a gas turns into a liquid or solid.

constellation A pattern of stars that people imagine as a recognisable object.

convection The movement of hot gas or liquid towards cooler regions, causing currents to flow.

core The dense central region of a planet, star or galaxy.

crater The hole blasted onto the surface of a moon or planet by the impact of a comet or other object.

crust The cool, solid layer of a moon or planet, lying on top of a hotter mantle or core.

dark energy The name for the mysterious "antigravity" force that may be causing the accelerating universe.

degrees (°) A measure of angles: one degree equals $\frac{1}{360}$ of a complete circle; ninety degrees equals $\frac{1}{4}$ of a circle.

diameter The distance across the middle of a circle or a sphere, measuring how large it is.

dust Carbon and other solid elements blown into space by ageing stars form dark clouds of what astronomers refer to as "dust".

dwarf planet Any world in our Solar System large enough to be round, but too small to dominate its region of space and clear away other small worlds. Dwarf planets reside with thousands of other worlds in the asteroid and Kuiper belts.

eclipse When one object, or its shadow, hides another—such as when the Moon hides the Sun, or Earth's shadow darkens the Moon.

electron A negatively charged particle that usually orbits the nucleus of an atom. Free-flowing electrons form currents of electricity.

EMU Extravehicular Mobility Unit, the technical term for the spacesuits astronauts wear.

event horizon The invisible boundary around a black hole that represents the point of no return—travel beyond the event horizon and there is no escape from the black hole's gravity.

extrasolar planet A planet orbiting another star.

galaxy A giant collection of billions of stars, often with many clouds of dust and gas.

gas giant A planet, usually very large like Jupiter and Saturn, made mostly of dense gas with no solid surface.

gravity The "force" exerted by all matter that attracts other matter and even energy like light. Gravity from the Sun keeps all the planets in orbit around the Sun.

greenhouse effect The process where carbon dioxide and other gases in an atmosphere trap heat. This makes a planet's surface warmer than it would be if it had no atmosphere.

helium An element made of two protons and two electrons. It is produced by thermonuclear fusion, combining hydrogen atoms inside stars.

Hubble Space Telescope Named for astronomer Edwin Hubble, this telescope orbits Earth. It can provide sharper views from space because it does not have to look through our blurry atmosphere.

hydrogen The simplest element, just one proton and one electron, that forms most of the universe.

infrared light Light waves too long for our eyes to see but that can be felt as heat.

Kuiper Belt object Small rocky and icy objects that orbit in a wide zone, beginning at the orbit of Pluto and extending much farther into space.

light-year The distance a beam of light travels in one year, equals 10 trillion kilometres (6 trillion mi).

mantle The middle layer of a moon or planet, between the crust and the core.

mass A measure of how much matter an object contains (Jupiter has more mass than Earth).

Messier An eighteenth century French astronomer who first catalogued many of the brightest star clusters and nebulas visible in the night sky.

meteor The name for the visible streak of light we see when a meteoroid burns up in our atmosphere.

meteorite The name for a small rock that survives its passage through an atmosphere and lands on a planet.

meteoroid The name for small bits of dust and rock orbiting the Sun, which can burn up in an atmosphere.

Milky Way galaxy The name for our own galaxy, home to our Sun and its Solar System.

moon Any world, from objects just a kilometre across to worlds thousands of kilometres in diameter, that orbits a planet.

nebula A "cloud" of gas and dust in the space between stars, which perhaps forms new stars, or may be blown into space by dying stars.

neutron A particle with no electrical charge found inside the nucleus (or core) of an atom.

neutron star A small, dense, collapsed star made of almost pure neutrons.

Oort Cloud A large, spherical region surrounding the Solar System but very far away, beyond Pluto and the Kuiper Belt. The Oort Cloud may be home to thousands of comets.

orbit The circular or elliptical path of a moon around a planet, a planet around a star or a star around the centre of a galaxy.

planet Any spherical world orbiting the Sun, or another star, large enough to dominate its area of its solar system.

planetary nebula A type of gas cloud blown into space by ageing stars, which cast off their outer layers of gas.

pole The end points of a world's axis of rotation, such as the North and South poles on Earth.

proto-galaxy An irregular collection of stars and nebulas present in the early universe, out of which familiar galaxies formed.

proton The particle with a positive electrical charge found inside the nucleus of an atom.

proto-planet A small newly formed world, orbiting a star, that may collide with other proto-planets to form larger, full-fledged planets.

radiation The term for any form of energy, such as radio, light, infrared energy or X-rays.

revolution The motion of a planet around its star (creating the length of the planet's year) or a moon around its planet (creating the planet's month).

rotation The motion of a world around its axis, creating the length of the world's day.

satellite An object that orbits a planet. A satellite is also a man-made space probe that orbits a planet or moon.

shepherd moon A small moon that orbits just inside or outside a ring, keeping the ring particles in place by the action of its gravity.

solar system The term for the Sun, or any star, and its family of planets and other objects.

space probe A robot craft sent from Earth to explore another planet or moon, or perhaps an asteroid or comet.

star A ball of gas large enough to shine by thermonuclear fusion.

supernova A star that explodes with such force it blows apart, leaving only a dense core that collapses to form a neutron star or black hole.

temperature A measure of how fast molecules and atoms are moving. We feel temperature as how hot or cold something is.

terrestrial planet A planet made mostly of rock, such as Earth or Mars.

thermonuclear fusion The process that occurs inside stars to combine light elements, such as hydrogen, into more complex elements, such as helium. This process releases energy as a by-product, allowing stars to shine.

ultraviolet light Light waves too short for our eyes to see but that, on Earth, can give sunburns.

universe The name for everything there is in space: all the galaxies, as well as the stars and planets in those galaxies. There is nothing bigger than the universe (at least not that we can see!).

vacuum The name for any volume with little air in it; space is a vacuum.

waning Moon When the Moon is going from Full to New, and its phase is shrinking each night.

waxing Moon When the Moon is going from New to Full, and its phase is growing each night.

X-rays Very powerful and energetic waves of radiation emitted by hot objects in space, such as the matter in accretion disks around black holes.

zero gravity The term, more correctly called "microgravity", which describes the weightless environment that astronauts experience as they orbit around Earth.

Index

Credits

The publisher thanks Alexandra Cooper for her contribution,
and Puddingburn for the index.

ILLUSTRATIONS
Front cover MBA Studios (main), Mark A. Garlick (supports);
back cover Mark A. Garlick
MBA Studios 32–3; **Karen Carr**© 11; **Mark A. Garlick** 4, 5, 6–7, 8–9, 10–11,
12–13, 14–15, 16–17, 18–19, 20–1 22–3, 24, 25, 34, 36–7, 38–9, 40–1, 42,
44, 45, 46, 48–9, 50–1, 52–3, 54–5, 56–7, 58–9, 60, 61, 62; **David Hardy**
60; **Mark A. Garlick/US Naval Observatory and Space Telescope Science
Institute** 22–3; **Steven Hobbs** 1, 3, 4, 5, 11, 23, 24–5, 26–7, 28–9, 30–1,
34–5, 42–3, 44–5, 46–7, 60, **Moonrunner Design** 33

PHOTOGRAPHS
Key t=top; l=left; r=right; tl=top left; tcl=top centre left; tc=top centre;
tcr=top centre right; tr=top right; cl=centre left; c=centre; cr=centre right;
b=bottom; bl=bottom left; bcl=bottom centre left; bc=bottom centre;
bcr=bottom centre right; br=bottom right

AAO=Anglo-Australian Observatory; APL/CBT=Australian Picture
Library/Corbis; ESA=European Space Agency; ESO=European Southern
Observatory; iS=istockphoto.com; N_ES=NASA/Earth from Space;
N_G=NASA/Great Images in NASA; N_H=NASA/Hubble Space Telescope;
N_J=NASA/Jet Propulsion Laboratory; N_N=NIX/NASA Image Exchange;
N_SF=NASA/Human Spaceflight; N_SP=NASA/Spitzer Space Telescope;

NASA=National Aeronautics and Space Administration; NOAO=National
Optical Astronomy Observatory; PL=photolibrary.com; SH=Shutterstock;
SOHO=Solar and Heliospheric Observatory

8cl N_ES **13**br iS **16**bcl PL **17**bcr N_H **19**bcr, br, cr, tcr, tr N_H **20**cl N_H
24bcl APL/CBT; bcl iS; br SH **25**bc N_H; bcl PL; bcr APL/CBT; br N_SP;
tr ESO **27**tr ESA **28**bc N_N; bcl, bl N_G **29**bcr N_G br N_J **30**bcl, bl, l N_SF
31br NASA **33**bcl, bcr, bl N_H **37**br APL/CBT; tcr SOHO; tr PL **41**tcl PL **43**br, tr
N_J; tcl ESA **44**bcl, br N_J **45**tcr, tr N_J **46**cl N_J **51**br N_A **52**bl NOAO; tl N_H
54bl Alan Dyer **56**tcr, tr AAO; tl N_H **58**tl PL **61**bc N_G; bcl, br, cr N_J; cl N_SF

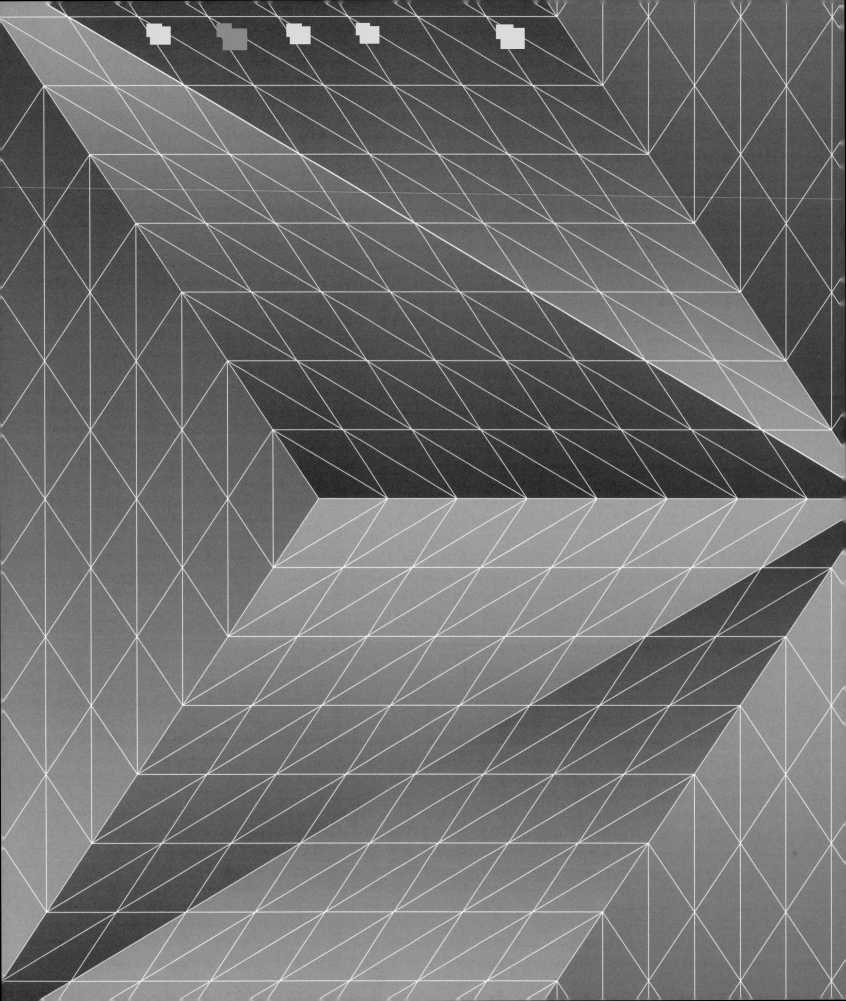